AGRICULTURE IN THE GATT

University of Plymouth Library

Subject to status this item may be renewed
via your Voyager account

http://voyager.plymouth.ac.uk

Exeter tel: (01392) 475049
Exmouth tel: (01395) 255331
Plymouth tel: (01752) 232323

338·184 JOS
0333658191

Agriculture in the GATT

Timothy E. Josling
Professor
Food Research Institute
Stanford University

Stefan Tangermann
Professor of Agricultural Economics
University of Göttingen
Germany

and

T. K. Warley
formerly Professor of Agricultural Economics
University of Guelph
Canada

First published in Great Britain 1996 by
MACMILLAN PRESS LTD
Houndmills, Basingstoke, Hampshire RG21 6XS
and London
Companies and representatives
throughout the world

A catalogue record for this book is available
from the British Library.

ISBN 0–333–65819–1

First published in the United States of America 1996 by
ST. MARTIN'S PRESS, INC.,
Scholarly and Reference Division,
175 Fifth Avenue,
New York, N.Y. 10010

ISBN 0–312–16237–5

Library of Congress Cataloging-in-Publication Data
Josling, Timothy Edward.
Agriculture in the GATT / Timothy E. Josling, Stefan Tangermann,
and T. K. Warley
p.   cm.
Includes bibliographical references and index.
ISBN 0–312–16237–5
1. Tariff on farm produce.   2. Produce trade—Government policy.
3. Agriculture and state.   General Agreement on Tariffs and
Trade (Organization).   I. Tangermann, Stefan.   II. Warley, T. K.
(Thorald Keith), 1930–   .   III. Title.
HF2651.F27A244   1996
382'.41—dc20   96–2814
                                                              CIP

10   9   8   7   6   5   4   3   2   1
05   04   03   02   01   00   99   98   97   96

Printed and bound in Great Britain by
Antony Rowe Ltd, Chippenham, Wiltshire

Contents

Preface

For many years, the three of us have shared a common interest in the agricultural policies of the major developed countries and in the effects of these policies on world agricultural trade. We have met often at conferences, symposia and workshops on international agricultural trade and policy, and have followed as analysts, commentators and observers each of the three major GATT negotiations in which agricultural issues have been addressed – the Kennedy, Tokyo, and Uruguay Rounds.

The original motivation for this book was to chronicle the story of the agricultural negotiations in these successive rounds and thereby to trace the evolution of the treatment of agricultural trade in the GATT. We felt that the story of repeated attempts to bring trade in agricultural products more surely into a rule-based international trading system should be recorded, not only for those who work in the area of agricultural trade but also for those with interests in international economic and political relations, international business, and economic history.

In the summer of 1990, we journeyed to Geneva to meet with GATT officials and members of national delegations to learn more about the status of the negotiations on agriculture in the Uruguay Round. At that time there was still the possibility of an agreement by the end of December 1990, with an agricultural component which would have provided the final chapter to the story. In the event the writing of that chapter had to wait until the negotiations had concluded, over three years later.

Meanwhile it became increasingly clear that the Uruguay Round would so radically change the rules under which agricultural trade takes place that it could not be treated as just another 'round' of negotiations. Trade in agricultural goods would indeed – for the first time since the inception of the GATT – be brought fully within a system of multilateral rules and disciplines. Accordingly, the scope of the book was expanded to allow the resulting new rules to be discussed in more detail and their implications to be explored.

In structuring the book we have followed a chronological sequence – beginning with the creation of the GATT, continuing through the 1954–5 review of the General Agreement and the Haberler Report, and dealing in turn with the Dillon, Kennedy, and Tokyo Rounds and the period following that, when the agricultural trade situation deteriorated sharply. We devote separate chapters to the negotiations in the Uruguay Round and the results for agriculture embedded in the Uruguay Round Agreement. A final chapter ventures a look ahead at some emerging issues and at the completion of the reforms initiated by the Uruguay Round.

The perspective that we have taken is largely descriptive, avoiding both quantitative analysis and censorious judgements, though undoubtedly our own opinions on these issues are apparent. The reader, and the passage of time, will have to judge whether our enthusiasm for what has now been accomplished in bringing trade in agricultural products into the rule-based GATT/WTO system is justified.

Writing the book was professionally satisfying and as close to fun as a serious endeavour can be. Each author contributed both substantive chapters and a different perspective. All parts of the book were collectively revised and harmonized, though some stylistic differences undoubtedly remain. Senior authorship is not assigned, and our names are listed in alphabetical order.

Throughout the gestation period of this book, we have worked with colleagues in the International Agricultural Trade Research Consortium – an organization that brings together economists interested in international agricultural trade matters – in various study groups and meetings. Our appreciation is extended to these colleagues for the long and fruitful discussions we have shared over the years of the issues surrounding the GATT negotiations on agriculture. We view the book as a contribution to the mission of the Consortium, and we hope that it will stimulate studies on other aspects of trade policy in agricultural goods.

During the preparation of the book we have been supported and encouraged by many people to whom we are deeply grateful. In particular, David Blandford, William J. Davey and Robert E. Hudec each read parts of an early draft and provided helpful comments. If we still managed to smuggle some errors into the manuscript, they should not be held responsible. In the final stages Harald Grethe provided invaluable technical support, Henning Twesten proficiently worked on the index, and Petra Geile skilfully brought the final manuscript into a proper order and acceptable format. Part of Stefan Tangermann's work on the book was generously supported by the Deutsche Forschungsgemeinschaft.

Finally, to our wives Anthea, Gabi, and Anita who sacrificed many weekends and evenings of our company to the completion of the book, we owe our lasting gratitude. To them we dedicate this, the fruit of our labours.

List of Abbreviations

AMS	Aggregate Measurement of Support
AoA	Agreement on Agriculture
APEC	Asian-Pacific Economic Cooperation Process
CAP	Common Agricultural Policy
CCC	Commodity Credit Corporation
CSE	Consumer Subsidy Equivalent
CET	common external tariff
CITES	Convention on International Trade in Endangered Species of Wild Fauna and Flora
CoA	Committee on Agriculture
CSD	Consultative Subcommittee on Surplus Disposal
CTA	Committee on Trade in Agriculture
EAI	Enterprise for the Americas Initiative
EC	European Community
ECOSOC	Economic and Social Council of the United Nations
EEA	European Economic Area
EEC	European Economic Community
EEP	Export Enhancement Program
EFTA	European Free Trade Association
EU	European Union
FAC	Food Aid Convention
FAO	United Nations Food and Agriculture Organization
FOGS	Functioning of the GATT System
FTAA	Free-Trade Area of the Americas
GNG	Group of Negotiations on Goods
GNG5	Negotiating Group on Agriculture
GNS	Group of Negotiations on Services
HS	Harmonized System
ICA	Intergovernmental Commodity Agreement
ICCICA	Interim Coordinating Committee for International Commodity Arrangements
IDA	International Dairy Arrangement
IDPC	International Dairy Products Council
IGA	International Grains Arrangement
IMC	International Meat Council
IMF	International Monetary Fund

ix

ITO	International Trade Organization
IWA	International Wheat Agreement
IWC	International Wheat Council
LTA	Long-term Textiles Agreement
MDS	*montant de soutien*
MFA	Multifibre Arrangement
MFN	most-favoured nation
MTN	Multilateral Trade Negotiations
NAFTA	North American Free Trade Agreement
NTM	non-tariff measures
OECD	Organization for Economic Cooperation and Development
OEEC	Organization for European Economic Cooperation
OMA	orderly marketing agreements
OPEC	Organization of Petroleum Exporting Countries
PPM	Processes and Production Methods
PSE	Producer Subsidy Equivalent
RTA	regional trade agreements
SACA	Special Agreement on Commodity Arrangements
SMU	Support Measurement Unit
SPS	sanitary and phytosanitary measures
SSP	Special Safeguard Provision
TAFTA	Trans-atlantic Free-Trade Agreement
TNC	Trade Negotiations Committee
TRIMs	Trade Related Investment Measures
TRIPS	Trade Related Intellectual Property Rights
TRQ	tariff-rate quotas
UNCTAD	United Nations Conference on Trade and Development
UR	Uruguay Round
VER	voluntary export restraints
WTC	Wheat Trade Convention
WTO	World Trade Organization

1 The GATT's Origins and Early Years

1 PLANNING FOR POSTWAR ECONOMIC COOPERATION[1]

It is one of the more comforting facts of modern history that a group of remarkable politicians, public servants, and academics in the United States and the United Kingdom began planning for postwar reconstruction and economic cooperation in the earliest years of the Second World War.[2] All were convinced that inappropriate international economic policies immediately after the First World War, and economic warfare in the 1920s and 1930s, had delayed postwar recovery, caused the Great Depression, and created the conditions which led to World War II. Insistence by the allies on reparations had impoverished and alienated Germany and sown the seeds of totalitarianism and aggression. Trade restrictions and discrimination, and exchange rate instability and competitive devaluations, had deepened the economic contraction. Economic nationalism and 'beggar-thy-neighbour' policies only served to produce further economic deterioration and political hostility. International conferences had acknowledged the problems and their causes, reaffirmed liberal principles, and indicated the directions in which solutions lay, but they had failed to curb economic warfare and stem the rising tide of political hostility.

There was a determination to apply the lessons of the past to the economic problems that it was anticipated would exist at the end of hostilities. Assistance with the reconstruction of all war-torn economies would replace demands for payment of reparations by the vanquished to the victors. International economic cooperation would be practised in place of confrontation. Provision would be made for cooperation on both monetary matters and trade since many of the pre-war trade restrictions had been introduced in response to balance of payments crises. Effective international institutions would be created within which nations could collaborate to avert the expected recession, assure the needed expansion and, beyond that, cooperate in the creation of a new international economic order which would be the basis of security and peace.

1

The two principal allies and economic powers, the United States and the United Kingdom, took the lead. It was clear that the United States would emerge from the war as an economic superpower and the world leader in reshaping international economic relations. US authorities were convinced that the creation of a strong international economy was the foundation of a lasting peace, and that only the US would have the resources and the commitment to play the leadership role. Building on the beliefs that had been the basis for the Reciprocal Trade Agreements Act of 1934, an open and expanding world economy was needed to absorb direct US foreign investments and the rising volume of exports needed to assure full employment and economic growth. The United Kingdom had entered the war as a major economic power, but as the war progressed it became apparent that the country would emerge from the conflict with damaged industries, weakened infrastructure, depleted monetary reserves, liquidated overseas investments, accumulated foreign debts, and severe balance of payments difficulties due to a merchandise trade imbalance and the contraction of the traditional surplus in invisibles. Though a period of economic management would be required at the conclusion of hostilities, it was clear that, for Britain too in the longer term, an open, predictable and expanding world economy was essential for economic recovery at home and for the maintenance of its ability to play a leading role in world affairs. Thus, both the US and the UK shared the objective of promoting economic recovery and development by concerting their international efforts to expand commerce, promote monetary stability, encourage investment and coordinate national economic policies. And both anticipated that these goals would be best accomplished by the creation of a set of permanent international institutions within which cooperation on global monetary arrangements and trade could be effected.

Organized thinking about these matters was well under way in London and Washington even before the US entered the war. Future trade policy was included in the Atlantic Charter in 1941 and the US–UK Mutual Aid Agreement signed in 1942. Discussions under this 'Lend–Lease' Agreement were held in Washington in September and October of 1943. Papers on future cooperation in both monetary arrangements and trade policy had been prepared. The discussions on financial cooperation laid the groundwork for the 1944 Bretton Woods agreement. On trade, the US Department of State had prepared a paper on a multilateral convention on commercial policy and the United Kingdom a paper on a commercial union. There was a broad measure of agreement at the expert level on the need for a multilateral convention on commercial policy and for an international institution that would supervise its implementation and resolve disputes between countries that subscribed to it. It was agreed that the rules of the convention should be made as precise as possible. There was also initial agreement on many of the substantive trade rules themselves. Tariffs had to be bound and substantially reduced. Discrimination in trade had to be outlawed. Quantitative restrictions were to be prohibited except where used – in a closely circumscribed manner – to deal with balance of payments emergencies. Export subsidies were not to be used. State trading entities were to behave like private traders. These objectives were to be enshrined in legally binding international obligations. It was recognized in the discussion that there was a relationship between trade policy and employment and that the convention would have to deal with intergovernmental agreements on commodities and with the trade effects of the actions of private cartels.

Though there was a meeting of minds amongst the working-level trade policy officials of both countries and their academic advisors, at the governmental level deep differences persisted between US and UK authorities on the relationship between liberal trade arrangements and the maintenance of full employment, on the abolition of quantitative restrictions, on the elimination of 'Imperial Preference', on the formula for cutting tariffs, and on several other matters. These differences had been present in the earlier discussions on the Atlantic Charter and the Lend–Lease Agreement, and they continued after the 1943 'seminar'[3] among experts.

Bilateral discussions continued as the war proceeded. Those on monetary policy and development finance progressed quickly. The multilateral Bretton Woods agreement establishing the International Monetary Fund (IMF) and the International Bank for Reconstruction and Development (later the World Bank and its affiliates) was signed in 1944. The discussions on trade policy went more slowly and it was December 1945 before the United States was ready to publish *Proposals for Consideration by an International Conference on Trade and Employment* (US Department of State, 1945).

Though this was put forward by the US, the British government signified its agreement with the essentials of the document as a basis for discussion. The central proposal was that there be created an International Trade Organization (ITO) which would oversee the operation of a multilateral code of trade conduct. The ITO was to complete the triumvirate of institutions through which international economic cooperation would be effected.

2 THE HAVANA CHARTER

At its first session, in February 1946, on a resolution tabled by the United States, the newly created Economic and Social Council of the United Nations (ECOSOC) agreed to convene a 'United Nations Conference on Trade and Employment' (UN, 1946a). The purposes of the conference were to be twofold, to agree on a charter for the ITO and to make a start on reducing tariff barriers. A Preparatory Committee representing 18 governments[4] was appointed to prepare a draft charter for eventual consideration at a plenary conference. The formal drafting began in October 1946 on the basis of a *Suggested Charter for the International Trade Organization* (US Department of State, 1946) submitted by the US. The *Suggested Charter* was an elaboration of the earlier *Proposals*, and, like these, had provisions on employment, development, restrictive business practices, and commodity policy as well as on commercial policy. The suggestions on trade policy were laid out in the chapter on commercial policy. It dealt with most forms of trade restriction. The language used in this chapter drew extensively on the language which had come to be used in the numerous bilateral trade agreements that had been concluded by the United States with various countries under its Reciprocal Trade Agreements Act of 1934.[5]

The negotiation of the ITO Charter was to take two years from the ECOSOC resolution in February 1946 to the signing of the Final Act of the United Nations Conference on Trade and Employment in Havana, Cuba, on 24 March 1948. The Charter went through three drafts in two meetings of the full Preparatory Committee in London (October–November 1946) and Geneva (May–August 1947) and a meeting of a technical drafting committee at Lake Success, New York (January–February 1947). The draft of a charter that emerged from the Geneva meeting of the Preparatory Committee was the basic document considered at the Havana Conference. The final act of the Havana Conference establishing the ITO is known as the Havana Charter.[6]

It was not to be expected that countries that differed in their stages of economic development, in the strength of their economies at the end of the war, in their political and economic systems, in their legal traditions, in their views of the role of the state and of the nature of international relationships, would easily agree to the draft ITO charter. In principle, US authorities wanted to craft a trading system that would ensure the efficient use of global resources, reduce the influence of governments on economic affairs, and provide for the enforcement of a tightly drawn and uniformly applied code of trade law. Other governments, including Britain, placed greater priority on such objectives as full employment, social goals, external payments balance, protecting infant industries, preserving trade preferences, and meeting the special needs of primary producers. Some had political philosophies that required an active role for the state both within national boundaries and at them. And many held the view that the purpose of an international agency was to clarify the common interest and mediate differences between states rather than to act as a law enforcement agency. Specifically on trade, there were also profound differences between countries on what should be the content of a multilateral code of trade conduct and on the institutional arrangements by which observance of the code should be assured.

These fundamental differences underlay the intense negotiations in the preparatory meetings and at Havana on the charter that was to be given to the ITO.[7] In the drafting process in London, New York, and Geneva and in the negotiations at Havana, the *Suggested Charter* was gradually transformed into the Havana Charter by the interaction of fundamental and particular differences between the two leading powers, the United States and the United Kingdom, and between these countries and the 52 other nations that met in Havana to define the shape of postwar international economic relations.

The agreement of a majority could only be reached by making compromises on almost every point of principle. By the time that the Final Act of the Havana Conference was reached, the pure vision of an open, nondiscriminatory, juridical trading system envisioned by the US and UK experts in 1943 had taken on a very different appearance in the chapter on commercial policy in the Havana Charter (Chapter IV). Important exceptions had been made in respect of existing trade preferences, and the creation of new preferences was permitted through the establishment of customs unions and free-trade areas. The use of quantitative restrictions was permitted to correct balance of payments problems, to promote development, and to support agricultural programmes. Both domestic and export subsidies were allowed. Elements of what would later be called 'special and differential treatment' were introduced into the Charter at the insistence of the developing countries (which were in a numerical majority at the Conference). State trading entities were subject only to weak rules on discrimination and transparency and an exhortation to behave like private traders. Broadly drawn clauses allowed countries to escape from their obligations if domestic industries were injured by trade liberalization. International commodity market management was provided for in Chapter VI of the

Charter. As Dam put it, 'The result was a grotesquely complicated document that included a multitude of compromises and that all too often saw a free-trade principle followed immediately by an exception authorizing trade restrictions' (Dam, 1970, p. 14).

The Charter for the ITO that emerged from Havana in the spring of 1948 had few friends. In particular, it failed to satisfy important groups in the two countries responsible for its conception. The British were preoccupied with a deepening national financial crisis rather than with international cooperation, and they had come to see that the United States would be unwilling to make cuts in its tariff schedule (commonly known as the Smoot–Hawley tariff) large enough to make it worth contemplating ending the Imperial Preference system. More important, in the United States, protectionists and isolationists joined economic liberals in complaining that the compromises made to reach agreement in virtually all the key areas had produced an outcome in which 'the exception . . . had devoured the rule' (Gardner, 1980, p. 282). The provisions on employment and development seemed to endorse state planning. Tariff preferences had not been eliminated. The exception for the continued use of quantitative restrictions in pursuit of balance of payment and economic development objectives threatened to continue discriminatory restrictions against US exports. Provision was made for the management of commodity markets. Foreign investment seemed to have less security than previously.

In the end, because the Havana Charter failed to meet US trade policy goals in the areas of preferences and the use of quantitative restrictions, and because Congress was hostile to the notion of surrendering authority on trade matters to an international institution, in December 1950 the US Administration quietly abandoned its efforts to secure Congressional approval for the charter for the ITO. Because other countries had made their ratification of the Final Act of the Havana Conference conditional on its acceptance by the United States, they too did not proceed. The ITO was stillborn. It would be April 1994 before agreement was reached on the creation of a permanent multilateral agency to oversee world trade, the World Trade Organization.

3   THE GENERAL AGREEMENT ON TARIFFS AND TRADE

The *Suggested Charter* had anticipated an early attack on trade barriers. In 1945 the US Congress had renewed the Reciprocal Trade Agreements Act. Since the Administration was anxious to act on this authority before it expired in June 1948, the US government indicated in December 1945 its desire to enter into negotiations with other countries for the reduction of tariffs and other trade barriers. A formal invitation to meet for this purpose was made to other countries at the London drafting conference in 1946. Thus, when the Preparatory Committee held its second full meeting in Geneva in 1947, its members were acting in two distinct capacities, preparing the draft ITO charter and the agenda for the planned world conference on trade and employment and, independently, as participants in tariff negotiations.

The technical drafting committee which had met at Lake Success, New York, through January and February 1947 had *inter alia* prepared a first draft of an agreement on tariffs and related aspects of trade conduct. Thus, the 1947 Geneva trade conference had two elements, reaching agreement on 'the substantial reduction of tariffs and the elimination of trade preferences' and on 'Procedures for Giving Effect to Certain Provisions of the Charter of the International Trade Organization by means of a General Agreement on Tariffs and Trade Among the Members of the Preparatory Committee' (UN, 1947a). The tariff negotiations and the second full meeting of the Preparatory Committee took place concurrently.

The tariff negotiations (and the Geneva Draft of the ITO Charter) were completed on 30 October 1947. The trade accord – a General Agreement on Tariffs and Trade, signed by 23 countries – consisted of the agreed tariff schedules (Part I) and (Parts II and III) the trade rules of such parts of the draft ITO Charter dealing with trade in goods *as it then stood* as were customarily placed in bilateral trade agreements to protect the agreed tariff reductions and bindings against nullification and impairment by national policy changes in other areas, and to settle disputes that might arise between the signatories.

The form of the negotiations, and the content and institutional characteristics of the Agreement, were a direct result of the US negotiators' limited authority under the Reciprocal Trade Agreements Act and the imperatives of acknowledging the gathering hostility of Congress to what was going on in Havana by avoiding giving 'the appearance of sneaking the ITO into effect by the back door' (Hudec, 1975, p. 46). The accord was framed as an executive agreement between governments. Care was taken to avoid any suggestion that the GATT was a new international organization. The negotiations were confined to tariff barriers, and were conducted on an item-by-item basis, between principal suppliers, and observing reciprocity. The bilateral agreements were then extended to all participating countries according to the nondiscrimination principle now enshrined in the most-favoured nation rule[8] (Article I). Recognizing that some parts of the Agreement might conflict with existing national legislation, countries participating in the tariff negotiations limited their legal obligation by agreeing to apply the General Agreement under a 'Protocol of Provisional Application', under which the signatories only committed themselves to apply the code of trade conduct contained in Part II of the Agreement 'to the fullest extent not inconsistent with existing [national] legislation'. Because it was anticipated that the General Agreement would be absorbed into the work of the proposed ITO, GATT was not provided with its own secretariat and budget. The business of the Agreement was to be conducted by the 'CONTRACTING PARTIES' acting 'jointly'. The secretariat services needed were provided by persons borrowed from an Interim Commission of the International Trade Organization which had been set up to prepare the work programme that would follow the Havana Conference. GATT's financing was similarly improvised, and remained so for many years.

Thus did the commercial policy provisions of the grand design for an International Trade Organization fortuitously survive in the General Agreement on Tariffs and Trade. In later years, notably at the ninth (review) session in 1954–5, after the demise of the proposal to create an ITO, attempts would be made to complete what many countries saw as the unfinished business of the Havana Charter negotiations. Proposals were made to create a permanent and more adequately administered trade institution (the Organization for Trade Cooperation), to write the commodity policy provisions of the ITO Charter into the General Agreement, and to substantively clarify and revise it. These attempts generally failed. The features of the agreement that arose as a consequence of its being an interim measure, devised in the context of a larger accord that did not materialize – its limited mandate, qualified legal obligations, rudimentary dispute settlement mechanisms, improvised institutional arrangements and, as we shall see, its unsatisfactory arrangements for agricultural trade – were to stay with the GATT for the next 47 years.[9]

4 THE ESSENTIALS OF THE AGREEMENT

Despite its unusual origins and partial mandate, the GATT has proved to be one of the more influential and respected of the postwar international institutions. Though conceived as a temporary arrangement and devoid of elements of the grander design discussed at Havana, the Agreement embodied a set of core beliefs held by major countries about how international commerce should be organized and conducted. And the content represented a comprehensive code of trade conduct. Behind the dense legal language of its articles and addenda and interpretive notes, lies the set of essential beliefs and principles that is the coherent foundation of the Agreement.[10]

(i)   People will enjoy a higher standard of living if they have free access to all the world's material and human resources through trade. The lowering and removal of barriers to trade should therefore be an objective for all governments, though they are not obligated to undertake it.

(ii)   All countries benefit from the order, fairness, and predictability that come from trade relations based on rules rather than on economic weight alone. The principles and standards of acceptable conduct that guide international commerce should therefore be codified and embodied in international commitments of a legal character.

(iii)   Market forces should be the major regulator of international commerce. Governments should be discouraged from erecting direct or indirect barriers to trade, and their interference with market forces should be permitted only in exceptional and specified circumstances.

(iv)   If national efforts to promote economic and social objectives affect trade relations, they are of international concern. Countries should respect the legitimate trade interests of other countries in devising and implementing domestic policies and programmes, and should seek to avoid harming them.

(v)   Where protection is given it should be afforded only by duties and similar direct charges. The tariff is a visible barrier that readily lends itself to negotiation. Indirect methods of protection should not be used save in exceptional circumstances and according to prescribed rules. (Most of the provisions of the Agreement are designed to protect the integrity of members' tariff concessions against nullification or impairment through such measures as quantitative restrictions, subsidies, dumping, countervailing duties, customs valuation and formalities, state trading, and government procurement.)

(vi)   Trade barriers of all types should be administered on a nondiscriminatory basis. Imports should be treated in the same manner

as national products in respect of charges and regulations. Any
trade benefit conferred on one country should be extended un-
conditionally to all other suppliers. New preferences should not
be created by raising trade barriers, and all preferences should
be progressively reduced and ultimately eliminated through the
liberalization of trade.

(vii) When negotiations to reduce trade barriers are entered into, they
should be carried out in such a manner as to establish a balance
of mutual advantage.

(viii) If the balance of advantage established in negotiations is subsequently disturbed ('nullified or impaired') by the action of a
country, means should be provided for its restoration. If a mutual accommodation cannot be reached by consultation, balance
can be restored by the measured release of the injured countries
from some of their obligations.

(ix) Maximum use should be made of the opportunities for consultation and cooperation provided by the GATT for making the above
principles effective, for solving common problems, and for resolving trade disputes.

The first three of these points constitute the foundational beliefs that
underlie the GATT; the remaining objectives are expressed operationally in the articles of the Agreement itself. The key principle and rules
of the GATT are the granting of nondiscriminatory treatment of imported goods from all sources at the border (Article I), and the treatment of imports in the same manner as domestically produced goods
once they cross it (Article III). The signatories to the Agreement also
have expectations which, though less explicit, are no less important.
These include mutuality of benefits, balance in the concessions that
lead to an agreed degree of openness of market access, reciprocity of
obligations, and a commitment to make trade more liberal and disciplined over time.

Throughout the history of the GATT, the degree of commitment to
these principles has unquestionably been influenced by countries' political and economic philosophies and stage of economic development.
And, from time to time, or persistently, most countries have failed to
conform to the principles or to fulfil their obligations in respect of
trade in the products of particular sectors. Agriculture is a prime example.
Nonetheless, the principles have provided a firm foundation for commercial relations between countries throughout the second half of this
century.

5 AGRICULTURE IN THE HAVANA CHARTER AND THE GENERAL AGREEMENT

Agricultural trade created problems for the architects of the international trading system from the earliest days of planning for the postwar era. In the informal discussions of 1943 and 1944 no special provision had been made for agricultural trade, though it was anticipated that there might be a role for commodity agreements to resolve some of the problems of trade in 'primary products'. However, by the time bilateral US–UK consultations on the content of a charter for the ITO were begun in London in the autumn of 1945, it was apparent that the trade rules would have to acknowledge the existence of extensive government involvement in national agricultural industries and the trade arrangements that typically accompanied national farm programmes.

In some countries agricultural policies had deep roots reaching back over a century. In others, agricultural price and income support policies had been introduced in the interwar years in response to the Depression and the collapse of agricultural incomes. During the Second World War most countries had engaged in regulation of their agrifood sector through price controls, production planning, allocation of production inputs, food rationing, and control of trade in farm and food products. It was likely that such measures would continue for some time after the end of the war to promote agricultural recovery, to offset the anticipated slump in farm product prices, and to extend to farmers the provisions of 'the welfare state' that was everywhere being created. By 1945 it was clear that it could not be anticipated that trade in farm and food products could soon be exposed to global competition and that special provisions would have to be made for trade in the products of that sector.

These provisions took the form of explicit exemptions from the general rules on the use of quantitative restrictions and export subsidies for trade in agricultural products. These two 'agricultural exceptions' were designed to accommodate the agricultural support programmes of the economically advanced countries. In addition, in order to take account of the distinctive problems of trade in primary products – which were of particular concern to developing countries – the possibility of there being a special regime for trade in commodities was anticipated.

Quantitative Restrictions

A key objective of the postwar foreign economic policy of the United States was to eliminate the use of quantitative trade restrictions. The use of such restrictions had proliferated during the interwar period and they were regarded as the epitome of protectionism, bilateralism, discrimination, and the subordination of market forces in international commerce to government decisions.[11] The 1943 bilateral discussion had anticipated that, in principle, quantitative restrictions would be prohibited. But Britain insisted that they be permitted for dealing with external payments imbalances for a transitional period after the war. By 1945 it was accepted that exceptions would be needed for three main purposes:[12] to cope with balance of payments difficulties; to protect infant industries in poor countries; and to accommodate certain agricultural programmes. These three exceptions were aimed, respectively, at the countries of Western Europe, the developing countries, and the United States.

The major agricultural exception for the use of quantitative restrictions (Article XI:2(c)(i)) was tailored to the requirements of the agricultural policies of the United States.[13] Under the Agricultural Adjustment Act of 1933 as extended and amended, in return for participating in supply control programmes, US producers of 'basic crops' were provided with support prices through 'non-recourse loans' offered by the Commodity Credit Corporation (CCC). The support prices were related to a 1910–14 'parity ratio' and were well above world levels. The accumulated stocks of the CCC were disposed of through domestic and foreign 'diversion' programmes. The feasibility of the US farm income support programme depended upon there being appropriate trade arrangements. Specifically, import controls were needed to prevent imports from undercutting domestic support prices and adding to the CCC's stocks. A 1935 amendment to Section 22 of the 1933 Act stated that whenever there was a danger that imports of agricultural products would 'render ineffective' or 'materially interfere with' any programme under the Act, then the President 'shall . . . impose such fees . . . or such quantitative limitations on any article or articles . . . as he finds . . . to be necessary'.[14] Accordingly, US authorities had insisted that provision be made in the *Proposals* for the legitimization of the use of quantitative import restrictions where these were needed to make domestic farm programmes effective.

In language that changed little, and over the gathering protests of most other countries, most notably the non-subsidizing smaller and medium-sized agricultural exporters, this first agricultural exception was carried into the *Suggested Charter*, into the commercial policy chapter of the Havana Charter, and into Article XI:2 of the General Agreement. Paragraph 1 of the Article states the general rule:

no prohibitions or restrictions other than duties, taxes or other charges, whether made effective through quotas, import or export licenses or other measures shall be instituted or maintained . . . on the importation . . . or exportation of any product. . . .

Paragraph 2 spells out the agricultural exception:

The provisions of paragraph 1 of this article shall not extend to . . . (a) restrictions temporarily applied to prevent or relieve critical shortages of foodstuffs . . .; (b) . . . restrictions necessary to the application of standards . . .; (c) . . . import restrictions on any agricultural product, imported in any form, . . . necessary to the enforcement of governmental measures which operate . . . (i) . . . to restrict the quantities of the like domestic product permitted to be marketed or produced . . . or (ii) to remove a temporary surplus. . . .

A number of safeguards were written into Article XI:2(c)(i) establishing the circumstances in which quantitative restrictions might be used, the conditions to be met, and the criteria to be applied. These were intended to prevent the exception's being used in a protectionist manner by preventing an increase in protection from existing levels.[15] First, imports could not be restricted unless the output or marketing of the domestic product was also restricted. Second, governments using the article must give advance public notice of their intentions to restrict imports and (under other articles)[16] be prepared to consult with countries that complain. Third, and most importantly, domestic output or sales had to be restricted to the same degree as imports. This safeguard was provided by the requirement that import restrictions

shall not be such as will reduce the total of imports relative to the total of domestic production, as compared with the proportion which might reasonably be expected to rule . . . in the absence of such restrictions . . . pay[ing] due regard to the proportion prevailing during a previous representative period and to . . . special factors . . . affecting trade. . . .

Additional interpretative notes were intended to narrow the range of 'like' products that might be subject to import restrictions, and to ensure that changes in relative production efficiency as between domestic and foreign producers were among the 'special factors' that were taken into account in allocating market shares.

This agricultural exemption to the general prohibition against the use of quantitative import restrictions was greatly resented from the inception of the General Agreement. It set a bad precedent from the outset for the treatment of agricultural trade under the GATT. And the subsequent unwillingness of the US to tailor even its agricultural import policies to the modest requirements of Article XI was soon to lead to the 1955 waiver, which cast a further pall over trade in agriculture. The exemption was to remain a source of contention between the member countries for almost half a century. It would be 1994 before it was finally removed by the agreement to 'tariffy' non-tariff agricultural trade restrictions reached in the Uruguay Round.

Subsidies

The architects of postwar economic cooperation were well aware that a comprehensive trade agreement would have to address the question of subsidies since production and export subsidies affected trade patterns and the competitive relationships between suppliers. And if not subjected to multilateral disciplines, national subsidies might be used to nullify the trade benefits that were expected to flow from agreements to reduce tariffs. Moreover, if countries were to give up some autonomy over the exchange rate of their currency under the arrangements for monetary cooperation supervised by the IMF, it could be anticipated that those with overvalued currencies would be tempted to use subsidies to maintain reserves and employment.

The regulation of trade-affecting subsidies on 'primary products' was subject to particular problems. International cooperation in the commodity field was considered a better alternative than subsidy competition, but past experience suggested that it would be difficult to negotiate durable intergovernmental commodity agreements. Subsidies of all types were especially prevalent in developed country agriculture. Many of these had some legitimacy in so far as they were concerned with such defensible objectives as rural income improvement, rural rehabilitation and development, and improving food supplies and food security. Their trade effects were frequently incidental. Where this was not the case, farmers could be expected to resist any reduction in the degree of subsidization they enjoyed, and their political influence would likely lead to their protests being heeded.

Export subsidies were generally regarded as a particularly repugnant trade measure. Since they were usually paid on only a fraction of total output they could be very large in unit terms. They were more readily available to wealthy countries than to the poor. They amounted to a unilateral nullification of the balance of benefits established in trade negotiations. They took markets away from more efficient domestic suppliers and foreign competitors in a highly visible way, and were likely to lead to subsidy escalation and confrontational trade relations.

Given this complex situation, it was clear that subsidies could not be banned, even in principle. Instead, a system of subsidy regulation would have to be devised that focused on controlling and mitigating their trade effects. It was felt, however, that a stronger line could be taken with subsidies on exports and manufacturing, on the one hand, as opposed to 'domestic' subsidies and subsidies on 'primary products', on the other.

The United States had difficulty with any proposals to govern subsidies on 'primary products', for US legislation provided subsidies to agriculture. Farm product prices were subsidized through the Commodity Credit Corporation's support purchasing operations (though the trade effects of these operations were not readily apparent). By contrast, the subsidies needed to export the farm products purchased by the CCC at support prices that were well above those prevailing in world markets were highly visible. They had been much resented by competing exporters in the pre-war years, and they would be resisted if paid in the postwar period. Since a postwar collapse of agricultural markets was widely anticipated, it was likely that they would be needed.

The regime for regulating subsidies put forward in the *Suggested Charter* made a generic provision for all subsidies and a separate provision for export subsidies. Within the latter, a distinction was made between export subsidies on primary products and other products. This differentiation was carried through the drafts of the ITO charter and into the Havana Charter itself.

The general obligation on subsidies contained in the Havana Charter required that a signatory that granted (domestic or export) subsidies, which were not production- and trade-neutral, and which therefore 'operate, directly or indirectly, to reduce imports or to increase exports . . .', must notify the other members of the character of the subsidy, its extent, the reasons for its adoption, and its trade effects. If the subsidy caused or threatened harm to another member, the country offering the subsidy could be called into consultation and be required to discuss the possibility that it might be modified. This general provision was carried into Article XVI of the GATT in 1947.

In keeping with the widely shared hostility to export subsidies, the Geneva draft of the Havana Charter would have required an end to the use of export subsidies on nonagricultural products after the Charter had been in force for three years. However, subject to specified conditions, member countries would have been permitted to continue to provide subsidies to the export of 'primary products'. The conditions set out in the Havana Charter were that the preferred solution of intergovernmental commodity arrangement had been attempted and failed; that the ITO had given its permission for export subsidies to be used; that the permission would be given only for a limited period; and that the subsidy would not result in the subsidizing country acquiring 'more than an equitable share of world trade . . .' in the product concerned.

The United States believed that, in practice, the ITO would never give permission for it to use agricultural export subsidies, and US negotiators knew that Congress would not tolerate the loss in sovereignty represented by the provision requiring prior international approval for the US to provide subsidies to move farm products into export markets. Furthermore, US negotiators argued that since the draft subsidies section of the Charter dealt lightly with production subsidies and stabilization schemes, they could not defend at home tight restrictions on the only price support action the United States could take in the event that burdensome surpluses developed. US negotiators also pointed out that the Reciprocal Trade Agreements Act did not authorize them to enter into a commitment on export subsidies.

For these reasons, at US insistence, only parts of the elaborate system for regulating subsidies, particularly export subsidies, were carried over from the drafts of the Havana Charter into the General Agreement. The general prohibition against the payment of export subsidies on nonagricultural products and the special regime governing the use of export subsidies for primary products that had been in the *Suggested Charter*, and which were simultaneously in the process of being written into the ITO Charter, were left out of the first working draft of the GATT prepared (by the US delegation) for the drafting committee of the Preparatory Committee, which met at Lake Success in January and February of 1947. They remained excluded from the General Agreement completed at the end of October that year, and from the revisions to the Agreement made in 1948 after the conclusion of the Havana Conference. And so the GATT started life with only the weak obligation to notify and consult on subsidies contained in what became Section A, paragraph 1 of Article XVI.

That this was a highly unsatisfactory situation was well understood, but little could be done as long as the United States was determined to prevent the GATT from disciplining its own agricultural support policies. It was only when the agricultural subsidy practices of others started to harm US agricultural export interests seriously that the United States changed tack and mounted a major assault on the adverse trade effects of agricultural subsidies. This was to occur with the formation of the European Economic Community in 1956. The story of agriculture in the GATT after 1956 is in no small part about attempts to bring the support provided by agricultural subsidies under effective GATT disciplines.

International Commodity Trade

There is a long history of the international community giving special attention to trade in what are variously called 'primary commodities', 'primary products' or simply 'commodities'. The League of Nations had sponsored conferences on the topic in the 1930s, and there was an even longer history of attempts to form commodity cartels. Hence, it was to be expected that international commodity trade would receive particular attention in planning for postwar economic cooperation. The particular problems to which commodities were prone were a tendency towards the development of persistent disequilibrium between supply and demand, the accumulation of burdensome surpluses, chronic price instability, and long periods of unsatisfactory returns to producers. The inelasticity of both supply and demand could make it difficult for these problems to be solved by the operation of normal market forces. In these circumstances, it was anticipated that international cooperation might be appropriate.

The *Proposals* and *Suggested Charter* issued by the United States envisioned that, in particular circumstances, there might be a need for a special regime of intergovernmental arrangements for trade in commodities. This provision was carried through the preliminary drafts of the charter for the ITO and into a separate chapter in the Havana Charter (Chapter VI) on Intergovernmental Commodity Agreement (ICA).

However, the view of the United States and other major countries was that government commodity cartels were to be regarded as temporary exceptions to the general rules of commercial policy, to be permitted only in specified circumstances, for a limited period, to enable particular objectives to be attained, and then only when subject to special rules and requirements. Though contested at every stage, this cautious approach to the approval and regulation of ICAs prevailed throughout the negotiation of the Havana Charter for the ITO. The objectives of the most restrictive ICAs – the 'control agreements' that might regulate production, trade, stocks and prices – were carefully specified in the Havana Charter as being to ameliorate serious economic difficulties that may arise for countries from problems in their commodity trade, to provide a framework and a time frame for the development and implementation of economic adjustments to underlying market conditions, and to prevent or moderate pronounced fluctuations in commodity prices. They could only be created for commodities that had characteristics that were found in the production and sale of a few agricultural products.[17] These were that small producers should account for a substantial share of output, that conditions of surplus should exist or threaten, that the surplus should cause hardship, that low price elasticities should prevail, and that there should be an expectation that the hardship could not be averted by the normal operation of market forces.

Additional safeguards were provided in the Havana Charter against abuse. ICAs were to be permitted only for 'primary commodities', narrowly defined (it was thought) as 'any product of farm, forest or fishery, or any mineral, in its natural form or which has undergone such processing as is customarily required to prepare it for marketing in substantial volume in international trade'. Agreements were to be sanctioned only if their usefulness had been established by Study Groups and Commodity Conferences. Their membership was to be open. They were to be concluded only for five years at a time. Equal voting weight had to be given to producer and consumer interests. The commodity schemes were to fall within the jurisdiction of the ITO, rather than the United Nations Food and Agriculture Organization (FAO), which, it was suspected, might give undue weight to producer interests and be inclined to see ICAs as a permanent feature of world trade in agricultural products. The objectives had to include the expansion of consumption and the assurance of supplies at prices in keeping with the agreed stabilization objectives. Most importantly, the price stabilization objective was specified as being

> to prevent or moderate pronounced fluctuations in the price of a primary commodity with a view to achieving a reasonable degree of stability on a basis of such prices as are fair to consumers and provide a reasonable return to producers, having regard to the desirability of securing long-term equilibrium between the forces of supply and demand . . . (Article 57)

All attempts at Havana to extend the range of products that might be subject to control schemes, and to endorse more ambitious pricing objectives, were rebuffed. In 1947, the Economic and Social Council of the United Nations (ECOSOC) had endorsed the principles on commodity agreements contained in the draft charter for the ITO and established the Interim Coordinating Committee for International Commodity Arrangements (ICCICA) as a UN agency responsible for facilitating action on commodity problems. More particularly, the ECOSOC resolution endorsed the US vision of the role of ICAs by stating of control schemes, in a resolution proposed by the US, that

> a fundamental principle of all schemes should be that, save in exceptional circumstances, they should not make the average price over a period higher or lower than it would otherwise have been. Their objective should be merely to reduce fluctuations around the long-term trend. (UN, 1947c)

This was a blow to those countries that wanted international trade in commodities to be managed in ways that would provide 'adequate' prices for primary product exports and thereby redistribute global income and promote development.

The General Agreement on Tariffs and Trade was not intended to be a comprehensive trade agreement – that was the purpose of the Havana Charter which was expected to follow. Accordingly, the first draft of the Agreement made no reference to commodity trade at all, and the General Agreement adopted in 1947 went no further than recognizing, in Article XX, as one of the several 'general exceptions' to the code of trade conduct, the right to use restrictive trade measures in pursuance of obligations under ICAs so long as the ICAs conform to internationally agreed principles, that is, those set out in the ITO Charter's draft chapter on commodity agreements and the ECOSOC resolution.

When the ITO failed to come into existence, the developing countries were left in the very unsatisfactory position of not being able to use the GATT as a vehicle for promoting the creation of commodity agreements. An attempt was made to add the essential provisions of Chapter VI of the Havana Charter to the GATT at the 1954–5 Review Session but it failed (see Chapter 2). Instead, the language of Article XX(h) was changed to reaffirm that trade measures required by international commodity agreements were permissible exceptions:

> nothing in this Agreement shall ... prevent the adoption ... of measures ... undertaken in pursuance of obligations under any intergovernmental commodity agreement which conforms to criteria submitted to the CONTRACTING PARTIES and not disapproved by them.

With GATT reflecting the view of countries that thought ICAs aberrant, those countries that favoured the creation of commodity agreements turned to United Nations agencies to further their cause. A permanent advisory Commission on Commodity Trade was created in 1954 and set alongside the ICCICA. A decade later the second United Nations Conference on Trade and Development (UNCTAD) was convened. The harnessing of trade to development in that institution was a return to a theme of the Havana Charter, and, in the 1970s, the UNCTAD's 'integrated programme for commodities' embodied objectives that the developing country commodity exporters had failed to achieve at Havana in 1947 and in the GATT in 1954–5.

The record of the CONTRACTING PARTIES' actions under the General Agreement with respect to intergovernmental commodity agreements might well have reflected disenchantment with the record of practical experience with ICAs. They have been few in number, fragile, and the source of continuous dissension amongst their signatories. But the negative stance towards this particular trade instrument is also consistent with the philosophy of the Agreement, the rules of which are designed to allow international production and exchange to be determined by market mechanisms, by diminishing and removing the interferences that prevent the play of market forces from determining the volume and direction of exports and imports. This view was closely attuned to the political and economic systems of the major industrial powers that wrote and operated the GATT, and especially of the United States. From the perspective of the architects of the Agreement, the close regulation by intergovernmental agreement of production, prices and market shares would be the very antithesis of the trade model the Agreement sought to create – an aberration to be tolerated for brief periods, in exceptional circumstances, and with strict safeguards, but not a system to be promoted as a continuing and integral part of the trading system. For this reason, though the GATT as written in 1947 genuflected towards the possibility that there might be a role for intergovernmental commodity agreements, it was not to be anticipated that the Agreement would be used to promote activity in this field.

2 Early Encounters: 1948–60

1 INTRODUCTION

In the late 1940s, rich and poor exporters of primary products, including temperate and tropical agricultural products, faced the present realities and future prospects in the trading system with great discontent and mounting apprehension.

The GATT had been explicitly written to accommodate the agricultural import controls and export subsidies of the United States. The production subsidies of the United Kingdom had been condoned, and its use of discriminatory bulk purchases and tariff preferences had not been effectively constrained. Continental European agricultural importers had secured authorization for the import quotas and foreign exchange controls they were using for balance of payments reasons. Agricultural production and export subsidies and quantitative restrictions were to be permitted for an unknown future period, subject only to weak multilateral surveillance and restraints. Conversely, little enthusiasm had been shown in the negotiations on the GATT and the ITO Charter for the intergovernmental commodity agreements which to some exporters of agricultural products were the preferred means of attaining short-run stability and long-term growth in their export earnings.

On every side, the low-cost exporters were confronted by the trade effects of the pervasive national agricultural policies of importing and exporting countries affluent enough to protect and subsidize their agricultural sectors. The reasons for agricultural support were varied, but expansion for balance of payments and food security reasons, and the acceptance by government of an obligation to offer farm people stable and adequate incomes, were the most common. Whatever the motivation, most systems of support had the common features of encouraging high-cost output and thereby reducing net import requirements or increasing net export availabilities. As a result, efficient agricultural exporters found the markets for their products reduced and destabilized. Food importing countries employed foreign exchange controls, import quotas, bulk purchase arrangements, and statist importing agencies. The major exporter of temperate zone agricultural products, the United States, was subsidizing agricultural production and using export subsidies and concessional sales to dispose on world markets supplies in excess of

21

domestic demand at regulated prices. Exporters who could afford the cost were drawn into competitive subsidization. With national markets insulated, world market conditions were determined by the interaction of the policy-determined excess supplies of exporters and excess demands of importers (Josling, 1977a). To add insult to injury, the resultant international distortions and instabilities were then cited as reasons for perpetuating and even extending national interventions!

Underlying the agricultural exporters' discontent with trade conditions and practices was the expectation that commodity prices would collapse in the early postwar years as they had at the end of the First World War. It was feared that agricultural subsidization and protection would exacerbate what were anyway expected to be difficult economic times.

As it became clear that the ITO would not be created, the developed and developing exporters were left to rely primarily on the General Agreement as a means of resolving conflicts on agricultural trade conditions and practices. Having failed to have written into the GATT the provisions they sought for their agricultural and commodity trade, they knew that the Agreement was deficient in content, legal precision, and standing as a basis for pressing their claim for fair and liberal trade in agriculture, and for the rule of law in this area of international commerce. But the Agreement was all they had, and they appealed to it immediately, repeatedly, and with vigour in the resolution of agricultural trade disputes, and they maintained continuous pressure to have the provisions of the Agreement changed in ways that would improve the performance of the world's agricultural production and trading system.

The story of these early encounters in the GATT on agricultural trade policy is summarized in this chapter. The period covered is from the inception of the General Agreement to the eve of the Dillon Round, 1948 to 1960.

2 THE EARLY ROUNDS

The contracting parties to the GATT engaged in formal trade negotiations in 1947, 1949, 1951 and 1956. These four early negotiating 'rounds' (as they came to be called) did little to improve the conditions of agricultural trade.

The negotiations conducted in Geneva in the period April to October 1947 had two purposes: the establishment of the General Agreement and the exchange of tariff concessions between the countries that were the original signatories to the GATT. On both scores the Geneva conference was a success. The GATT was signed, and bindings and reductions were made on some 45 000 tariffs. However, since agricultural trade was even then distorted mainly by non-tariff measures, the binding and reduction of tariffs left the major impediments to agricultural trade intact. Where tariffs on farm and food products did exist, their protectionist intent, links to national agricultural policy objectives, and political sensitivity resulted in few tariffs being bound or reduced, or only small cuts from high levels.

Much to the disappointment of the United States, for which the elimination of discrimination in the trading system was a major goal of commercial policy, the major system of trade preferences then in existence – that operated by Britain and its present and former colonies – was little affected by the Geneva Round. The United Kingdom was reluctant to abandon 'Imperial Preference' altogether, and willing to reduce it only in return for a bigger reduction in the US tariff than US negotiators were authorized to offer. The US (Smoot–Hawley) tariff schedule established in 1930 had been set so high that even the full 50 per cent cut authorized by Congress for the 1947 negotiations would have left US tariffs on many items at prohibitive levels.

The second GATT trade conference was held at Annecy, France, in April 1949. It was mainly concerned with admitting ten new members and establishing their 'entry price' through bindings and reductions in their tariff schedules. The original members did not conduct tariff negotiations among themselves. As at Geneva two years earlier, the non-tariff trade barriers maintained by the newcomers were not addressed, and few concessions were made on the protective tariffs they maintained on imports of farm and food products. Ominously, even as the GATT negotiations were in progress in Annecy, the 81st US Congress was debating farm legislation that would eventually set minimum price supports for basic commodities at high and rigid levels (90 per cent of parity for 1950 and 80 per cent for 1951). Worse still, Congress was debating an amendment to Section 22 of the Agricultural Adjustment Act which would have asserted the primacy of US national farm legislation over that country's international obligations under the GATT.

An equally important action that showed the tension between agricultural and trade policy objectives in the United States occurred shortly after the conclusion of the Annecy Round. In that Round the US had reduced its tariff on butter from 14 to 7 cents a pound and established a tariff quota of 60 million pounds at that rate. However, the Commodity Credit Corporation had acquired a stock of butter under the dairy price support programme, and the Second War Measures Act required that imports be restricted whenever that was essential to the orderly disposal of government stocks. Accordingly, the US Secretary of Agriculture was forced by law to place an embargo on imports of foreign butter. This was the first of several agricultural policy developments in the United States that would eventually lead the US to seek a waiver from its obligations under the General Agreement on its agricultural trade.

The 1949 Annecy Round, and accompanying developments in US agricultural and trade policies, were a disappointment to those exporters who hoped to see agricultural trade liberalized. They were equally discouraging to those countries that wished to see world commodity trade organized and managed. During the Annecy Session a proposal had been tabled suggesting that Chapter VI of the Havana Charter, which dealt with intergovernmental commodity agreements, be ratified in advance of the ratification of the whole Charter. This proposal was unacceptable to the United States and nothing came of it.

The third GATT round was held in Torquay, England, in the winter of 1950–1. It too was focused on tariff reductions and so had limited relevance to, and yielded few results for, agricultural trade. Furthermore, it was in this round that the limits became apparent to the results that could be expected from a negotiating method that entailed reciprocal bargaining between principal suppliers on an item-by-item basis. But no change in tariff negotiating technique could help the agricultural exporters when tariffs were not the major trade impediment and the tariffs that existed were not offered for binding and reduction.

Agricultural exporters' expectations received a further setback at Torquay when, during the course of the negotiations, President Truman announced that the Administration was abandoning its attempts to persuade Congress to ratify the Havana Charter establishing the ITO. This dashed the hopes both of those who had looked to stern Charter rules to discipline subsidies and quantitative restrictions, and of those who had hoped to see the Charter used to encourage the creation of ICAs.

The fourth GATT negotiating round was held at Geneva in 1956. It followed the Review Session held in late 1954 and early 1955. As is discussed below, at the Review Session numerous proposals to discipline and expand agricultural trade had been considered, and for the most part rejected. Accordingly, it was not to be expected that the delegates would make progress in the negotiating conference held a few months later. Protectionist pressures were running strongly in the

United States at the time, and there was widespread disillusion in the US with the GATT as an institution and as a vehicle for creating a liberal international trading system. Few of the developing countries participated in the 1956 Geneva Round, and European participation was constrained by the fact that no solution to the problem of tariff disparities had been found. For all these reasons, little progress was made in any area in the 1956 Round.

3   THE RULES REVISITED: THE 1954–5 REVIEW OF THE GATT

The Review Session, which ran from October 1954 to March 1955, was an important event in the life of the General Agreement. In the seventh year of its existence, the GATT was only a provisionally applied executive agreement between governments. It contained only some of the trade rules of the commercial policy chapter of the Havana Charter. Though its workload was growing, it was still serviced by a small Secretariat temporarily on loan from the Interim Commission for the International Trade Organization. There were signs that the traditional item-by-item method of bargaining on tariff reductions had run its course. There was deep discontent with the balance of benefits provided by the Agreement. In particular, low-income countries were demanding that multilateral commercial policy be used to accelerate their economic development, while the medium-sized and smaller exporters of agricultural products were demanding that their comparative advantage be released by measures that would improve their access to the food importers' markets, and lower the level of trade-distorting agricultural subsidies. The purposes of the Review Session were, therefore, to give permanence to the trade institution and consolidate its procedures, to revise the GATT's rules and strengthen its disciplines, and to examine the possibility of adding areas that were in the Havana Charter but had been omitted from the GATT. In the area of the GATT's rules, the exceptions that permitted the use of quantitative import controls and export subsidies in agricultural trade were the most contentious. The subject of quantitative restrictions had been given new urgency by developments in US agricultural policy and trade behaviour. Export subsidies were being extensively employed in agricultural trade, and a closely related subject was that of the growth of concessional sales and donations under 'food aid' programmes. Of the subjects on which the Charter had said much but the GATT said little, the role of international commodity agreements was among the most pressing issues. All this meant that agricultural trade would occupy much of the Session's agenda.

The US Waiver

As noted earlier, the intellectual architects of postwar trade policy had wished to have a flat prohibition on the use of quantitative import restrictions, which were regarded as the antithesis of a liberal trade order since, by their nature, they provided absolute protection and entailed bilateralism, discrimination, and the direct regulation of private activity. However, a GATT report in 1950 had shown that the use of quantitative restrictions on agricultural imports was widespread (GATT, 1950). The smaller exporters' discontent was focused on their use by the United States.[1] The United States was making extensive use of quantitative controls on agricultural imports for protective purposes. US quotas on sugar dated back to 1934. Imports of cotton, wheat and wheat flour, and later butter and cheese were controlled under Section 22 of the Agricultural Adjustment Act of 1933. Rice, fats, and oils imports were restricted under the Second War Powers Act of 1942. Section 22 of the Agricultural Adjustment Act empowered the President to raise tariffs by 50 per cent or impose quotas on imports if these occurred or threatened to occur in such quantities as to render ineffective or materially interfere with any agricultural adjustment programme. The US could no doubt have claimed exemption for existing legislation under the Protocol of Provisional Application, and for some products (though not for dairy products, wheat, and cotton) it might have been claimed that effective national supply controls were in place. But whatever the legalities of US practice, the fact was that the leading advocate of trade liberalism was setting a bad example.

Worse was to follow. As Eric Wyndham White (the Executive Secretary to the GATT in its first 21 years) had noted, although Article XI was 'largely tailor made to United States requirements . . . the tailors cut the cloth too fine' (Dam, 1970, p. 260). Congress wanted to restrict agricultural imports still further in open violation of the GATT.

Section 22 of the Agricultural Adjustment Act contained a major safeguard against the use of the power to restrict agricultural imports to contravene United States general commercial policy goals. A 1948 amendment specified that 'No proclamation under this section shall be enforced in contravention of any treaty or other international agreement to which the United States is or hereafter becomes a party.'

However, in 1950, certain Senators attempted to reverse this situation. An amendment to Section 22 would have provided that 'No international agreement hereafter shall be entered into by the United States or renewed or extended or allowed to extend beyond its permissible termination date in contravention of this section'.

The amendment was finally eliminated under threat of a Presidential veto, but it was clear that Congress intended to reserve the unilateral right to restrict agricultural imports. And Congress got its way the following year when, in the course of passing the Trade Agreements Extension Act of 1951, Section 22 of the Agricultural Adjustment Act was amended to permit its use to impose quotas and duties on farm imports regardless of any international agreement entered into by the United States: 'No trade agreement or other international agreement heretofore or hereafter entered into by the United States shall be applied in a manner inconsistent with the requirements of this section.'

As Leddy noted, 'this is the only instance on record of deliberate action by Congress directing the President to violate trade agreement commitments' (Leddy, 1963, p. 208).

In 1950, in the course of the Korean War, the US Congress attached a rider (Section 104) to a bill extending the Defense Production Act which required the President to impose restrictions on imports of a wide range of agricultural products if they would impair or reduce domestic production, or interfere with the storage and marketing of the domestic product. Since restricting dairy product imports had little to do with the prosecution of the war in Korea, the intent of Section 104 was clearly protectionist. Acting under the requirement of this law, imports of cheese and casein were severely curtailed (albeit imports accounted for only 3 per cent of consumption) and imports of butter and milk powder were excluded. Thereafter, the Executive used both Section 22 of the Agricultural Adjustment Act and Section 104 of the Defense Production Act to impose quantitative import restrictions on a range of agricultural products in response to demands for restrictive action voiced by farm groups and Congress.

The protests of the dairy product exporting countries were immediate and vigorous, and sustained over the next three years.[2] The United States was charged with being in violation of its legal obligations, and did not deny it. In 1951 United States dairy import quotas were formally found to infringe Article XI.[3] At the seventh session in 1952 the Netherlands was allowed to retaliate (under the nullification and impairment provisions of Article XXIII) by restricting permitted US wheat flour imports to 60 000 metric tons (BISD 1S/33).[4] Section 104 of the

Defense Production Act was repealed in June 1953, but the dairy import quotas were continued under Section 22 of the Agricultural Adjustment Act. This was not an improvement in the position with respect to Article XI:2(c) since domestic production and marketing were still not being controlled at all, let alone to the same degree as imports.

Claiming that time was needed to resolve the inconsistencies between the dictates of its national farm legislation and its legal obligations under the General Agreement, at the 1954–5 Review Session the United States sought a waiver under Article XXV:5 from its commitments under the provisions of Article II (import fees) and XI (import quotas), in so far as such commitments might be regarded as being inconsistent with the action the Administration was obligated to take under Section 22 of the Agricultural Adjustment Act. The United States was thereby seeking to legalize its position. Small nations might be able to play fast and loose with the letter and spirit of the General Agreement, but 'great powers must be concerned . . . about the effects of their actions on the system as a whole' (Keohane and Nye, 1973, p. 127). The United States was not able to conform to the requirements of the General Agreement but the architect of the system could not simply ignore it without threatening to bring down the whole edifice.

The waiver was granted in March 1955 to the extent 'necessary to prevent a conflict' with Section 22 (BISD 3S/34–5). The other member countries had no choice but to accede to the request, for the alternative might have been the withdrawal of the United States from the GATT. The waiver was exceptionally broad. It embraced commodities other than the dairy products which had caused the complaints. It was open-ended in that it applied not only to existing programmes but also to any that might subsequently be introduced. The prior approval of the CONTRACTING PARTIES for new restrictions was not required. Unlike the agricultural waivers later granted to West Germany and Belgium, the US waiver was not limited in time. There was no legal obligation to adopt policy measures that would obviate the need for import fees and quotas. US negotiators successfully resisted all attempts by the working party that considered the request to circumscribe the waiver. Subject only to the requirement to report annually and to enter into Article XXIII consultations if requested, US agricultural import policy had effectively been removed from international legal constraints.

The importance of this development can hardly be exaggerated. Dam described it as 'a grave blow to GATT's prestige' (Dam, 1970, p. 260). Others observed that no other major country was prepared to abide by the GATT rules so long as the United States had its privileged position in agriculture. At the sixth review of the waiver it was stated that 'this action . . . had probably caused more serious damage to the fulfilment of the objectives of the General Agreement than any other single factor' (BISD 9S/261).

That others were using quantitative import restrictions, that the United States had at least regularized its position, that injury to other countries could be identified and compensated for, and that Section 22 was used with restraint, all gave small comfort. It was especially discouraging for the smaller exporters who needed US support if any progress was to be made in holding back the rising tide of agricultural protectionism in Europe and Japan. Furthermore, the retention of the waiver over the following forty years – despite the rebukes voiced at each year's review – did little to strengthen the credibility of the United States in its later attempts to seek the liberalization of trade in agriculture.

The 1955 United States agricultural waiver was to have a chilling effect on international agricultural trade policy. It continued as a highly visible reminder that at a crucial moment in the development of the Agreement, the United States gave primacy to its national agricultural interests over its international trade obligations, sought sanctions for the use of the very import quotas that its commercial diplomacy had sought to banish, and demonstrated that agricultural products were indeed subject to different trade rules.

European Hard Core Waivers

Whereas US import quotas for agriculture were the focus of attention in the early 1950s, after 1955 it was the quantitative import quotas of the West European countries that came under attack. The United States was not the only country using quotas to limit agricultural imports. Initially, many other countries had been able to claim GATT coverage for their import quotas under Article XII, the balance of payments exception to the use of quantitative restrictions. But this exception became more difficult to invoke as the 1950s progressed and balance of payments positions improved and exchange controls were lifted. Furthermore, in the first half of the 1950s, under the Code of Liberalization of the Organization for European Economic Cooperation (OEEC), quantitative restrictions were progressively relaxed and removed on most nonagricultural traded products. When quantitative restrictions were examined at the 1954–5 Review Session it was apparent that many European countries were maintaining quantitative import controls, despite the great improvements that had been achieved in their trade and payments situation. However, some of these countries claimed that intolerable social disruption and economic hardship would be experienced in some sectors if their import trade was abruptly liberalized. Accordingly, it was agreed that such countries should seek a waiver to permit, for a period, the retention of import quotas on the products of the sectors concerned. The decision recognized that

> some transitional measure of protection by means of quantitative restrictions may be required for a limited period to enable an industry having received incidental protection from those restrictions which were maintained during the period of balance-of-payments difficulties to adjust itself to the situation which would be created by removal of these restrictions. (BISD 3S/39)

Unlike the US waiver, which was granted on the same day, 5 March 1955, the conditions attached to what were termed 'hard core waivers' were explicit and strict. There was an obligation to demonstrate the social hardship claimed, and petitioning governments had to commit themselves to an acceptable plan and a timetable for actions that would remove, within five years, the need for the waiver granted for their 'residual restrictions'. This was a signal that quantitative restrictions were no longer to be ignored.

It soon became apparent that in Western Europe most of the 'residual restrictions' were for agricultural products. But because they did not wish to have to agree that their agricultural import quotas were protectionist, nor commit themselves to a plan and a timetable for their abandonment, countries were not willing to apply for hard core waivers for their residual quantitative restrictions on imports of agricultural products. In fact, only Belgium did so.

West Germany was the largest European food importer maintaining import quotas to protect its agriculture. When it was challenged at the Twelfth Session (October–November 1957), it first tried to defend its import practices on balance of payments grounds and then on the claim that the quotas were mandatory under its agricultural marketing laws, and were therefore covered under the protocol of accession it had signed in April 1951. When both these grounds were rejected, West Germany was forced to seek a waiver. This it did in 1959, but not before the West German delegate had observed that 'it has been more and more recognized that the pertinent provisions of the General Agreement relating to this important field are no longer realistic and need to be revised' (BISD 6S/65–6). West Germany chose not to seek a 'hard core' waiver[5] (BISD 8S/31–50) for its agricultural import practices because it was unwilling to give 'any undertakings or assurances . . . about future commercial policy as it affected agriculture'. No doubt many other countries, if pressed, would have had to say much the same, particularly those that, at the time, like West Germany, were turning their attention to the design of a common agricultural policy for the regional economic grouping taking shape in Europe.

The agricultural trade problem created by the use of quantitative restrictions by the continental European countries in the 1950s was never resolved. In the case of the European Community countries, it was overtaken by the problems created by the replacement of national import quotas with the Community's import levies. The other West European countries were to continue to use import quotas for agricultural products for another three decades. In most of these cases, quantitative import restrictions were 'grandfathered' either by the Protocol of Provisional Application or by the protocols of accession.

Subsidies

At the inception of the GATT, the use of subsidies in agriculture was subject only to the notification and consultation provisions of what became Part A of Article XVI. Not surprisingly, this weak obligation did nothing to constrain the proliferation of subsidies to agriculture. Discontent with the trade distortions caused by the pervasive and effectively unconstrained subsidization of agriculture mounted in the early 1950s. At the insistence of the smaller and developing agricultural exporters – which could not themselves afford to use production and export subsidies yet bore the brunt of their use by others – the whole subsidy issue was revisited at the 1954–5 Review Session.

It was agreed to strengthen Article XVI by incorporating the intent of the articles on export subsidies that had been in the Havana Charter but which had been omitted from the General Agreement in 1947. This was done by adding Part B, *Additional Provisions on Export Subsidies*. Part B reintroduced the Havana Charter prohibition on the use of export subsidies to manufactures. They were to be forbidden after 31 December 1957. But though the signatories agreed to 'seek to avoid' subsidizing exports of 'primary products', their legal right to use export subsidies for agricultural products was retained. This second agricultural exception was constrained only by the obligation not to use export subsidies 'in a manner which results in [the] contracting party having more than an equitable share of world trade . . . account being taken of the shares of the contracting parties . . . during a previous representative period, and any special factors which may have affected or may be affecting such trade in the product.' This was the only mild constraint on its agricultural export subsidy practices that the United States would support, and it resisted an attempt by other exporters to have the 'equitable share of world export trade' rule interpreted as applying to individual markets. Following the lead of the Havana Charter, an interpretive note attempted to limit the definition of 'primary product'. Part A, which was the unchanged single clause of the old article, was still the only constraint on domestic subsidies in agriculture.

Surplus Disposal

In addition to their preoccupation with the use of agricultural export subsidies as such, in the 1950s the smaller and developing exporters were also concerned by the practice of disposing of surplus stocks on highly concessional terms or as outright gifts. These stocks were the excess supplies acquired by public agencies under price and income support arrangements. Although several countries occasionally had made food aid available to war-ravaged European countries or to developing countries, the United States was the one country that operated a comprehensive and continuous programme of surplus disposal. In the early postwar years, US surplus commodities went mainly to Western European countries and Japan under Section 402 of the Mutual Security Act. Later, they were directed to developing countries as economic assistance and for famine relief, mainly under the Agricultural Trade Development and Assistance Act of 1954, better known as PL 480. Though much was made of the market development potential of this legislation, of the value of making food available to a hungry world, and of the resource transfers entailed in making food supplies available to poor countries on concessional terms, the motivation for the programme was to dispose of the excess supplies that were continuously acquired by the CCC under support programmes in the absence of price adjustments and effective production controls.

At first it was thought that surplus disposal would be a passing phenomenon, but by the mid-1950s it was apparent that it was an enduring feature of the world food trade system. Moreover, the scale of surplus disposal operations grew to a level where it was disrupting the trade of countries exporting on commercial terms. In the early 1950s, concessional sales and gifts accounted for 20–30 per cent of all United States agricultural exports and some 60 per cent of its wheat exports.

Attempts to bring national surplus disposal operations into a multilateral framework were made in the FAO with a proposal in 1946 for the creation of a World Food Board, a proposal for the establishment of an International Commodity Clearing House in 1949, and in 1954 an invitation from the UN General Assembly to the FAO to examine the feasibility of establishing a World Food Reserve. Nothing came of these initiatives. Eventually, this growing threat to the stability and fairness of the world agricultural trading system became a source of serious friction between exporters and it was taken to the GATT.

Australia took the initiative at the 1954–5 Review Session with a proposal to add a new article to the Agreement that would regulate the disposal of surplus stocks (BISD 3S/229). The article would have required prior notification of surplus disposal operations and an obligation to ensure that concessional supplies added to consumption without displacing commercial imports. It also provided for third-party arbitration in the case of disputes and proven damage. This proposal was unacceptable to the United States. Accordingly, the CONTRACTING PARTIES settled for the adoption of a resolution urging responsible behaviour (BISD 3S/50–1). This provided for consultations among principal suppliers on the orderly liquidation of surplus stocks. The expansion of consumption was to be encouraged 'where practical', as was the avoidance of prejudice to the interest of other countries. Donors were required to 'give sympathetic consideration' to the views of other contracting parties. It was also agreed that the GATT should conduct an annual review of surplus disposal operations. These reviews were conducted for some years, but gradually withered into the submission of annual summary reports.

Effectively, the multilateral regulation of the non-commercial disposal of agricultural surpluses was turned over to the FAO, a body whose resolutions carried no legally binding obligations. The FAO, in 1954, had adopted a code of conduct to be observed in the operation of surplus disposal programmes (FAO, 1954). The principles and guidelines that constituted the code were designed to protect the commercial interests of third countries. Like the GATT resolution of the following year, they called for prior notification, avoidance of market disruption, additivity in consumption in recipient countries, and consultation with countries whose interests were affected. The mechanism for overseeing the observance of these principles was the Consultative Subcommittee on Surplus Disposal (CSD), a subcommittee of the FAO's Committee on Commodity Problems. As time passed, the members of the CSD gave greater precision to the consultative process by classifying transaction types, devising a workable definition of 'surplus disposal of an agricultural commodity in international trade', agreeing on the measurement of the 'usual market requirements' that should be purchased commercially, and formalizing the procedures for bilateral consultations and multilateral evaluation of surplus disposal transactions.

The combination of code and institutional arrangement seems to have worked quite well. The essentials of the code were written into the 1958 revision of US PL 480 and incorporated into the World Food Programme of 1963. They would reappear in the GATT in the Kennedy Round, where a multilateral food aid commitment became part of the 1967 International Grains Arrangement. Much later, the subject surfaced a third time in the GATT in the Uruguay Round Agreement on Agriculture, where it became part of the new rules for export subsidies.

Commodity Agreements Revisited

In the 1950s, if those who wanted world trade in 'primary products' to be desubsidized and liberalized were to be disappointed, so too were those countries who wanted it to be organized and managed. As the agricultural exporters' problems of instability and inadequacy of prices and earnings mounted in the early 1950s, they continued to press the case for a larger role for ICAs. The subject was reopened at the Review Session when, on an initiative by some developing countries and smaller exporters of temperate zone agricultural products, a Working Party was established to examine proposals for inserting into the Agreement provisions along the lines of Chapter VI of the Havana Charter. The Working Party was to re-examine the principles and objectives of ICAs and prepare a draft international agreement which could provide a framework for possible multilateral action in the field of commodity trade.[6] The resultant Special Agreement on Commodity Arrangements (SACA) would have established an organizational structure for commodity matters, reaffirmed that ICAs were appropriate when conditions existed in commodity markets that could not be dealt with adequately by normal market forces, and spelled out a series of objectives that signatories might seek to achieve by joint action. The objectives were modest and market oriented. They did not call for ICAs to be used to redistribute world income, as had some of the proposals made at Havana. Even so, the report of the Working Party met opposition from the United States and West Germany. Nor did it attract strong support from the smaller agricultural exporters, which by this time were entertaining hopes that agricultural trade could be brought within the liberalizing disciplines of the GATT rather than relegated to a separate agreement.

The SACA was discussed at the tenth and eleventh regular Sessions in 1955 and 1956. However, at the eleventh Session, it was concluded that no agreement could be reached on general criteria for approving and operating ICAs (BISD 5S/87). Accordingly, having noted that other multilateral agencies also had responsibilities in the commodity trade field, the signatories resolved to keep commodity trade under review. This they did on the basis of annual reports on the situation and trends in commodity markets provided by the chairman of the ICCICA (who was nominated by the CONTRACTING PARTIES). Effectively, this was a rejection of a market management approach to agricultural trade problems within the GATT. This was far from the action desired by those suffering from agricultural trade problems, but the annual discussions of these problems and their causes allowed the frustrations and complaints of the exporters to be kept on the GATT agenda and in front of the contracting parties.

ICAs as such were to surface in the GATT in subsequent years. The Haberler Report in 1958 identified a role for buffer stock agreements in enhancing the stability of price and foreign exchange earnings for primary product exporters, and Committee III (established after the Haberler Committee had reported) spent a good deal of its energies on examining commodity trade problems. But ICAs were never again candidates for action in the way envisioned by some countries at the GATT's inception and in the 1954–5 Review.

4   THE HABERLER REPORT

There were substantial setbacks for the advocates of agricultural trade reform at the Review Session. Nor was the situation improved in the period immediately following. Late in 1955, at the tenth GATT session, Japan was able to protect its agricultural import quotas under its protocol of accession to the General Agreement. The following year, the 1956 Geneva Round did little to further liberalize overall world trade, and nothing for agricultural trade. The unease was exacerbated in 1956, when the process of forming the European Economic Community was begun and early indications appeared that the Six intended to embrace a protectionist and inward-looking common policy for agriculture. Anxieties were raised further by the prospect that the world's largest food importer, the United Kingdom, would apply for membership in the Community and radically change its food import arrangements. In all, there was a gathering sense of malaise in the international trading system in the mid-1950s, and much of this disenchantment was concerned with the bad and deteriorating conditions in trade in agriculture resulting from widespread agricultural protectionism.

All these considerations led to a decision at the twelfth Session in November 1957 to invite a group of independent experts to study trends in world trade and report their findings to the CONTRACTING PARTIES (BISD 6S/18). The expert group consisted of four distinguished international economists, Gottfried Haberler, Professor of Economics at Harvard University, James Meade, Professor of Political Economy at the University of Cambridge, Roberto de Oliveiro Campos, Professor of Economics at the University of Brazil, and Jan Tinbergen, Professor of Development Programming at the Netherlands Institute for Advanced Economic Studies in Rotterdam. Professor Haberler was chairman and the group came to be known as the Haberler Committee. Its first report was entitled *Trends in International Trade* (GATT, 1958). This landmark document is commonly called the Haberler Report.

The Haberler Committee was established to examine 'past and current international trade trends and their implications' and to address, in particular, the three issues of

the failure of the trade of the less-developed countries to develop as rapidly as that of industrialized countries, excessive short-term fluctuations in prices of primary products, and wide spread resort to agricultural protection. (BISD 6S/18)

The Committee was not asked for advice on possible action, and it was specifically enjoined not to comment on the policies of individual countries.

The Committee did a remarkable job in a few short months. It presented its report in October 1958, in time for it to be considered by the trade ministers assembled in Geneva for the thirteenth GATT session in October and November. The report covered all three areas included in the Committee's mandate, including agricultural trade.

On agriculture, the report did a number of things. First, it documented the pervasiveness of interventionist policies for agriculture in the developed countries and drew attention to the diversity of policy instruments that were employed. The latter included the full array of domestic subsidies on products and inputs, aids to exports, and import barriers. It was noted that most of the instruments were non-tariff measures, and that, as such, their use was but poorly controlled by the General Agreement. The description of agricultural policies was not in itself new, for agricultural policy reviews had been a regular feature of the work in the 1950s of the OEEC (later the Organization for Economic Cooperation and Development) (OEEC, 1956/7, 1960; OECD, 1961, 1967). What was new was the firm link drawn by the Haberler Committee between national farm income support programmes and the conditions of world agricultural trade.

Second, the Committee attempted to measure the extent of agricultural protection. Given the multiplicity of policy instruments which were used to transfer income to farmers directly and indirectly, the Committee recognized that the best measure of protection 'would be to measure the percentage by which the price (including any subsidy) received by the domestic producer exceeded the price at which the product was available from foreign suppliers or could be sold to foreign consumers' (GATT, 1958, p. 83–4). But having anticipated the need for what would later be called 'an aggregate measurement of support', the Committee settled for measuring the nominal rate of protection revealed by comparing world market and national support prices. These were generally high, and in some countries extraordinarily so.

Third, the Committee drew attention to the fact that a number of developing countries were also exporters of the agricultural products that were protected in rich countries, so that poor countries were harmed by the import protection, subsidy, and surplus disposal operations of rich countries.

The other major finding of the Committee was that, because of the Agreement's weak hold on agricultural protectionism, the agricultural exporters were not deriving the benefits from the operation of the Agreement that they had the right to expect. The Committee also warned that negative consequences for the trading system could flow from the creation of regional trading blocs such as the European Economic Community.

Because of its broad mandate, the recommendations of the Haberler Committee were similarly far-reaching. Some were sharply focused on the problems of trade in agriculture. Those recommendations that were so focused were explicit. Essentially, the Committee advised the countries of Western Europe and North America to lower the degree of protection accorded their agricultural producers; to permit freer trade in agricultural products; to make wider use of the deficiency payment as a farm income support measure (because of its transparency and avoidance of demand-side distortions); and to introduce programmes to assist agricultural structural reform and farm modernization.

In principle, this was sound advice. But with the United States clinging to its waiver and West Germany seeking one, with the EEC Six embarked upon the articulation of the Common Agricultural Policy (CAP), the time (late 1958) was not propitious for the advice to lead to radical action. Moreover, there was a sharp difference of views on the implications for trade policy of the Committee's findings and on the nature of the appropriate policy response.

5 COMMITTEE II

The immediate action that arose from the discussion of the Haberler Committee's report was the establishment of three further committees, known as Committees I, II and III. These committees were to work on a 'programme of action directed . . . towards . . . [an] expansion of international trade' (BISD 7S/28). This action programme would entail three elements: the trade problems of the developing countries, the trade problems caused by national agricultural policies, and a new GATT round of trade negotiations.

Committee I wrote reports on the possibilities of a new round of tariff negotiations. Its work led to the Dillon and Kennedy Rounds. Committee III addressed measures that would improve the position of the developing countries in the international trading system. A mark of its influence on the operation of the General Agreement was the addition of Part IV in 1964.

Committee II was directed to study 'problems arising out of the widespread use of non-tariff measures for the protection of agriculture, or in support of the maintenance of incomes of agricultural producers' (BISD 7S/28).

The work of Committee II had four components. First, it was to continue the work begun by the Haberler Committee in assembling data on the agricultural policies of the major industrialized countries, their use of non-tariff measures, and the external effects that flowed from these policies and their accompanying trade arrangements. But whereas Haberler's Committee had approached the subject on a general level, Committee II's enquiries were to be country and commodity specific. Second, Committee II was to provide a vehicle for multilateral consultations on the farm and food trade policies of individual countries. In practice, these consultations were used by the exporters to confront other countries about the likely trade effects of agricultural policy changes the latter were considering. Third, the Committee was asked to examine the adequacy of GATT rules for agricultural trade in light of the ubiquity of the national agricultural policies, which provided national producers with a high degree of protection and which had far-reaching effects on international trade in farm and food products. Finally, the work of the Committee was expected to lead to recommendations on steps that could be taken to improve the situation.

Committee II held a series of meetings over the next few years. It presented its results in a series of reports (BISD 8S/121–31, BISD 9S/110–20, and BISD 10S/125–67). Maintaining a pace that reflected the sense of urgency that had developed since Professor Haberler and his colleagues had reported, Committee II reviewed the agricultural policies of 34 countries. It also examined the problems in world markets for particular commodities and the way these were influenced by national agricultural and trade policies. Its detailed reports confirmed the Haberler Committee's general findings on the extent of the protection provided in the industrialized countries, and described the country-specific ways in which this was done. However, by far the most important consultations (and confrontations) held under Committee II's auspices were those that concerned the evolution of the Common Agricultural Policy of the European Economic Community (EEC). An account of the consultations with the EEC was published in a special report (GATT, 1962a).

Once the broad principles and concepts of the CAP were settled in January 1962, Committee II became the focal point of the exporters' complaints 'that the Community had concocted about as watertight a system of protection as could be devised'[7] and one that was bound to have adverse effects on their exports. To the exporters, the CAP epitomized the worst features of agricultural protectionism that the Haberler Committee had so recently described. They protested that the Six were creating an autonomous agricultural policy without regard to its external effects. Providing adequate farm incomes through the support of the market price of products would inevitably stimulate production while discouraging consumption. The variable import levy would provide absolute protection to Europe's farmers. Absent supply controls, and there was nothing to prevent the EEC's ceasing to be an importer of the regulated products and becoming an exporter. Should the Community develop exportable surpluses, it would be in a position to steal third markets from the commercial exporters with its proposed export 'restitutions'.

To all those protestations the Community's spokespersons offered reassurances. The EEC would follow a prudent price policy. The attainment of self-sufficiency was not a policy objective. They did not like export subsidies either. They would limit their use and avoid abuses.

The Community was aware of its international responsibilities. And the Six were well able to stonewall in this manner since all the malevolent external results of the CAP that were alleged would depend upon the level of common support prices that was set, and this matter had not yet been decided. Hence, the deliberations in Committee II could provide no basis for an unequivocal determination as to whether the Six were in violation of their obligations under the General Agreement. There was no outright condemnation of the CAP. The EEC countries never needed to secure a waiver for any part of it. No ruling on the legality of the variable levy system was sought. It was left for individual countries to reserve their rights under the Agreement and invoke them if they chose.

The recommendations of Committee II echoed those of the Haberler Committee in that they centred on a moderation of agricultural protection in both importing and exporting countries and the adoption of structural policies that would reduce the farm population. These recommendations were notable only in so far as they were made by representatives of national governments rather than by independent academics. The overall conclusion was more telling. In its third report the Committee drew the conclusion that

> there has been extensive resort to the use of non-tariff devices, whether or not in conformity with the General Agreement, which, in many cases, has impaired or nullified tariff concessions or other benefits which agricultural exporting countries expect to receive from the General Agreement. Hence the Committee concluded that the balance which countries consider they had a right to receive under the General Agreement has been disturbed. These developments are of such a character that either they have weakened or threaten to weaken the operation of the General Agreement as an instrument for the promotion of mutually advantageous trade. This situation raises the question as to the extent to which the GATT is an effective instrument for the promotion of such trade. (GATT, 1962b, p. 25)

As Kock observed, in GATT's legal terms and language, these were very strong words (Kock, 1969, p. 174). Committee II did not proceed to examine GATT rules as they applied to agricultural trade, and this part of its mandate was subsequently dropped. Thus, it was left to the trade ministers at their November 1961 meeting to decide what should be done. They resolved that the CONTRACTING PARTIES 'should adopt procedures designed to establish the basis for the negotiation of practical measures for the creation of acceptable conditions of access to world markets for agricultural commodities' (BISD 10S/27).

As a first step they proposed the establishment of study groups on cereals and meats. These were formed, and the Cereals Group held its first meeting in Geneva in February 1962. In the discussions at the sixteenth Session the previous November about what should be done about the agricultural protectionism documented in Committee II's reports, the Minister of Finance and Economic Affairs for France, M. Baumgartner, had aired a controversial plan for the organization of world trade in agricultural products. The 'Baumgartner Plan' (see Chapter 4) was one basis for discussion in the Cereals Group.

Looking back, the report of the Haberler Committee and the work and reports of Committee II made several contributions to the evolution of international agricultural trade policy. First and foremost, they established the links between national farm programmes and the conditions of international agricultural trade. This pointed to the conclusion that, if lasting and fundamental improvements were to be made, countries would have to negotiate about their domestic policies for agriculture as opposed to negotiating about the frontier measures that accompanied the internal programmes, or about attenuating their external effects as reflected in market shares, which was all that had been done under the GATT to that point. If acted upon, this would be a departure of primordial importance for the contracting parties and for the General Agreement. Second, the two Committees increased the transparency of agricultural policy domestically and internationally, and by so doing uncovered the most trade-disruptive instruments and countries. Finally, the Committees were the vehicle by which the United States and the 'non-subsidizing exporters' (as they would later call themselves), including the developing country exporters, were able to clarify and press their discontent and create the political momentum needed to make agricultural trade the centrepiece of the Dillon and Kennedy Rounds.

3 The Dillon Round[1]

1 BACKGROUND

The dominant political and economic event of the late 1950s was the signing of the Treaty of Rome on 25 March 1957 and the formation of the European Economic Community on 1 January 1958. This event was welcomed by the United States and other members of the international community on two grounds. First, the stimulus to economic growth that was anticipated to flow from European economic integration would provide a strong economic foundation for the Western military alliance. Second, and more importantly, it was thought that economic integration would prevent the re-emergence of the traditional political conflicts amongst the nations of Western Europe that had led to the wars which, three times in the previous century, had soaked Europe in blood.

But though desired on security and political grounds, the formation of the EEC posed several difficult economic problems for the rest of the world, and especially for the United States, the leader of the Western alliance. In the first place, there was the question of the legality of the European customs union under Article XXIV of the General Agreement which, subject to defined conditions, permitted the formation of customs unions and free-trade areas. Article XXIV had been written into the GATT in 1947 mainly to accommodate Benelux, an economic association of Belgium, the Netherlands, and Luxembourg.[2] The EEC was a regional preferential trading arrangement on a far vaster scale than had been anticipated by the drafters of the Agreement. Second, though the Rome Treaty was only the barest framework for the policies that would eventually be created by the six founding governments, its promise of a common agricultural policy, and of association agreements with overseas territories and countries enjoying preferential trade arrangements with the member states, raised the spectre of a deepening agricultural protectionism in Western Europe and the creation of new regional preferences in international trade. This threat had to be confronted.[3] Third, at a more mundane level, the phased replacement (beginning in January 1959) of the national tariffs of the member states by a common external tariff (CET) would entail changing tariff rates

42

that had been bound in earlier GATT rounds, thereby potentially disturbing the balance of concessions previously established. For all these reasons, it was necessary to engage the Community in negotiations.

At a practical level, it was necessary to ensure that the formation of the EEC met the requirements of the GATT Article XXIV:5(a) that the common restrictions were 'not on the whole . . . higher or more restrictive than the general incidence of the duties and regulations of commerce applicable in the constituent territories prior to the formation of such a union . . .' or, if they were, that appropriate compensation was paid. Also, there was a desire to minimize the potential for trade diversion by persuading the Community to lower its CET so that the degree of economic preference would be reduced. Finally, in the broadest political terms, there was a need for the rest of the world to reshape its commercial relationships with this important new entity, in ways that would ensure that the Community's trade and other policies conformed to the precepts of a liberal international economic order.

Other developments in Western Europe in the late 1950s reinforced the economic and political anxieties engendered by the union of the Six. The other major countries of Western Europe, led by Britain, had attempted to deal with the EEC by forming a Europe-wide free-trade area under the auspices of the Organization of European Economic Cooperation.[4] When this approach broke down late in 1958,[5] it was widely anticipated that Britain would apply for membership in the EEC, as would other European countries. This threatened to enlarge the area of discrimination and, from an agricultural perspective, take the world's largest importer of non-tropical agricultural products behind what was already feared would be an 'agricultural iron curtain'. Moreover, it was certain that the United Kingdom would seek preferential access to the enlarged Community for the Commonwealth suppliers of grains, dairy products, sugar, and other primary commodities.

The problems posed by the formation of the EEC and its potential enlargement were intertwined with the weighty issues that had been highlighted by the Haberler Committee in its report on trends in international trade. The formation of the EEC seemed to epitomize everything that Haberler and his colleagues had found objectionable in the conditions of international commerce. The creation of the European Economic Community was a massive assault on the principle of nondiscrimination. Its association agreements might alleviate the trade problems of a few associated developing countries but exacerbate those of a larger number. Its proposed common agricultural policy might extend agricultural protectionism. Its enlargement to embrace most of the countries of Western Europe would make the situation even worse.

The United States took the lead in attempting to influence the evolution of the commercial relationships of a uniting Europe with the rest of the world. Its aims were to lower the height of the impending CET and to moderate the protection conferred by the still-to-be-settled CAP. These achievements would anchor Europe in a liberal Atlantic alliance and minimize the trade diversion caused by the European preferences. It would also help the United States close the merchandise trade deficit it had begun to run by preserving access to the European market for US exports of industrial and agricultural products.

In 1958, the US Congress extended the Trade Agreements Act for four years until 30 June 1962. Using this authority, in 1959 D. Douglas Dillon, the US Under-Secretary of State for Economic Affairs, proposed to Ministers attending the thirteenth GATT Session that there be a two-stage trade negotiating conference. The first stage would be conducted under Article XXIV:6. It would be concerned with the rebalancing of concessions necessitated by the Community countries replacing previously bound national tariffs with the CET, and with binding the new rates. The second phase would be a normal tariff reduction conference.

This fifth round of GATT trade negotiations became known as the Dillon Round. The timing was right in the sense that there was the opportunity to influence Europe's decision-making in Brussels by agreements reached at the bargaining table in Geneva. In any event, the Six were scheduled to begin adapting their national tariffs to the CET on 1 January 1962, which meant that the Article XXIV:6 negotiations would have to be completed before that date, and before the US Administration's negotiating authority expired.

The EEC welcomed the proposed GATT round for several reasons. First, some member countries – notably West Germany and the Netherlands – were strong advocates of liberal trade, at least in industrial products. Second, the Six were anxious to allay the anxieties that had been generated by the creation of this large preferential trading area and to show that the formation of the Community would not harm third countries or the international trading system as a whole. Third, the Dillon Round would enable the Community to deal with the EFTA Seven on a multilateral basis and thereby disarm the Seven's demands for a bilateral deal. As an indicator of their good intentions, EEC spokespersons announced that for most products they would extend to other countries the first and second 10 per cent internal tariff cuts to be made on 1 January and 1 June 1959, provided these cuts did not result in the national tariffs falling below the anticipated CET rates. This offer excluded agricultural products and other sensitive items.

In 1959 the newly established Committee I reported on the possibilities of a new round of tariff negotiations, and made recommendations on the negotiations' topics, scope and modalities (BISD 8S/101–21). The negotiations began in Geneva in September 1960. The compensation phase ran until May 1961. The Dillon Round of negotiations on tariff reductions which followed immediately was concluded in March 1962. Both phases were concerned with adjusting and binding tariffs and, as such, were not focused on agriculture where protection was provided by non-tariff measures. Nonetheless, in adjusting the tariffs that did exist, some deeper problems of trade in agriculture were encountered.

2   THE TARIFF NEGOTIATIONS

In the compensation phase conducted under Article XXIV:6, the Community took the view that the rebinding of tariffs at CET rates required no net compensation since the adverse trade effects of increases in particular national tariffs, as member states aligned on CET rates, would be adequately compensated by reductions in the level of tariffs in other countries, and on other products, and by the stimulus to growth that would result from the integration of the economies of the Six.[6] This opinion was not shared by the countries negotiating with the Community. Each took the view that they should be compensated for the increase in the rate of protection they would face on specific products in the particular export markets in the Six that they had been able to develop. The Community stuck to its position and refused to permit the first phase of the negotiations to be conducted on a bilateral and product-by-product basis.

The tariff-reduction phase of the negotiations – the Dillon Round proper – was equally disappointing to the Community's trading partners. The negotiations at first seemed to hold promise. Except for some sensitive industrial products and agricultural products (other than some tropical products), the Community offered to make a linear cut of 20 per cent in its proposed common external tariffs. Congress had authorized the US Administration to reduce tariffs by 20 per cent of their level on 1 July 1959, or by two percentage points, or to cut any rate over 50 per cent ad valorem to that level. A 20 per cent all round cut would have been significant, and a linear cut in tariffs would have offered an escape from the limitations of the discredited product-by-product tariff bargaining technique.

However, the US Trade Agreements Extension Act of 1958 required item-by-item negotiations and observance of 'peril-point' tariff levels (established by the US Tariff Commission), below which tariffs might not be cut lest the reductions cause or threaten serious injury to US industries. Also, Congress had not authorized negotiations on the non-tariff trade barriers that protected some sections of US agriculture. With neither the United States nor the EEC willing to make offers that would improve access for the smaller and developing agricultural exporters, these countries were unwilling to offer reductions in their tariffs on imports of industrial products. This, in turn, caused the major countries to scale back their offers. Thus was set in train a process of retreat from the 20 per cent target.

In the end, 'after a year of whittling offers down to the lowest common denominator', the progressive unravelling of offers resulted in a situation whereby 'the unweighted average tariff on manufacturers may have been cut by one percentage point' (Curzon and Curzon, 1976, p. 174).

3 AGRICULTURE

Though some of the results subsequently turned out to be immensely important, at the time agriculture did not figure prominently in the reciprocal bargaining of the second phase of the Dillon Round. The EEC agreed to tariff reductions and new bindings at CET rates on some minor agricultural and processed food products for which protection was provided mainly by tariffs, and the United States requested that the tariff be bound at zero on some agricultural inputs that the EEC imported. Amongst the requests that the EEC granted was the binding of a zero duty on soya beans and soya bean meal and cake, and the binding at zero or low rates of duties on some feed ingredients, which were later to become known in the EEC as 'cereal substitutes'.[7] It is doubtful if the chief American and EEC agricultural negotiators (Raymond Iaones and Jean Marc Lucq) realized the significance of their decision at the time it was made. Certainly, no one predicted that EEC imports of US soya beans and products would soar from a value of $200 million in 1960 to a peak of $2.5 billion in 1988, and that EEC imports of cereal substitutes would increase from fewer than 2 million metric tons in the early sixties to more than 18 million metric tons in 1987. Nor was it anticipated that the growth of EEC imports of oil-seeds, oil-seed products, and cereal substitutes would so increase the Community's expenditures on its grains policy, that 30 years later its demand to be allowed to 'rebalance' protection between grains and non-grain feeds and oil-seeds would become one of the most contentious issues in negotiations on trade in agriculture in the Uruguay Round (see Chapter 8).

The crisis in agriculture occurred earlier, in the first or compensatory phase of the 1960–2 negotiations, when the EEC denounced the bound tariffs of its member states,[8] and declined to bind revised rates, on all products that were to be subject to 'market organization' under the still-to-be-settled common agricultural policy. The major commodities involved were food and feed grains, dairy products, sugar, and some meats, including poultry meat.

The first response of the United States was to ask the EEC to bind ceiling rates on the variable levies that would eventually replace the previously bound tariffs. Indeed, the United States threatened collapse of the negotiations if reasonable ceiling bindings on levies were not given. When this request was rebuffed, the United States asked instead that the EEC assure 'guaranteed access' to the European market for the most important commodities of which the United States was a large traditional supplier. This too was refused, although the EEC did offer assurances that the variable levy system would not be operated in such a way as to damage US exports.

Faced with EEC intransigence, the United States had to choose between allowing the Dillon Round to founder on its failure to negotiate constraints on the EEC's ability to impair access to the European market for US agricultural exports, or concluding the negotiations and thereby allowing the Six to proceed with the creation of their customs union for manufacturers and a common market for agriculture, and, beyond that, to broaden and deepen their economic integration and their military and political cooperation. On a decision by President Kennedy, acting on the advice of the Department of State,[9] the newly installed Administration chose the latter course. However, the retreat was regarded as only tactical and temporary, for the decision had already been made to engage the Community – and confront its common agricultural policy – in a major new set of multilateral trade negotiations of far greater scope and ambition than the Dillon Round.

Meantime, the United States sought to prevent the US–EEC agricultural situation from worsening. It did this by concluding two agreements with the Community, one for corn, soya beans, wheat, rice, and poultry, and a second more elaborate agreement on quality wheat.[10] Both agreements had two key provisions. First, they committed the Six to refrain from changing their import regimes in ways that would make them more restrictive in the interval before the CAP came into effect. This was a standstill agreement. Second, the Community agreed that when the CAP did come into effect, it would enter into negotiations with the United States (and under the agreement on hard wheat with other interested and signatory contracting parties), on the basis of the latter's 'unsatisfied negotiating rights' as of 1 September 1960, including the right to seek compensatory concessions or, failing that, to make retaliatory withdrawals of tariff concessions granted to the EEC. The agreement on hard wheat further committed the Community to take 'appropriate measures' to remedy any decline in wheat imports that was a consequence of the implementation of the CAP.

On this unsatisfactory note, the Dillon Round ended on 7 March 1962. The trade diversion potential of the EEC had not been lowered and existing restraints on Europe's proposed common agricultural policy had been denounced. The issues of the potential effects of the CAP on the international trading system for agricultural products as a whole, and the effects on other countries' concrete economic interests, had been left for future determination.

4 SIMULTANEOUS DEVELOPMENTS

While the Dillon Round was in progress in the period September 1960 to March 1962, there were several other developments that added to the mounting discontent with the condition of world trade in agriculture. The most important of these was the progressive unfolding of Europe's common agricultural policy. The Treaty of Rome had done little more than agree that there should be a common farm policy to replace the national agricultural policies of the individual member countries, and to state, in Article 39, that its objectives should include ensuring 'a fair standard of living for the agricultural population' and action 'to stabilize markets', while taking account of 'the particular character of agricultural activities arising from the social structure of agriculture and from structural and natural disparities between the various regions . . . [and] the need to make the appropriate adjustments gradually'. It was not difficult for exporters to see that this implied a continuation of the practice of bolstering farm incomes by the support of product prices. It was apparent that Europe's farmers were to be given a preferred position in selling in their own markets, absolute protection from outside competition, and a mechanism for dumping onto world markets supplies produced in excess of Community requirements.

Hence, the Dillon Round began with the EEC embarked on the construction of a common farm policy that would stimulate food autarky, and which had the potential not only for reducing export sales to Europe but also for generating exportable surpluses that could progressively displace commercial sales in third markets. Furthermore, denouncing existing GATT bindings cleared the way for the Community to follow this course should it so choose. However, nothing could be done in the Dillon Round itself to head off the threat, since the member States did not actually agree on the details of the internal price support and trade regimes for individual commodities (and on the broad arrangements for financing the common agricultural policy) until January 1962, only two months before the Round had to be concluded under US trade legislation. Furthermore, the common support prices – which would be the key to the CAP's trade effects – would not be determined until much later, and little could be done until these were known.[11]

The second discordant note for trade in agriculture was struck in the middle of the Dillon Round in the summer of 1961, when the United Kingdom applied for membership of the European Economic Community. This was a move that the United States welcomed on strategic grounds, and there were some who thought that Britain's membership would strengthen the Community's 'outward' orientation and temper the protectionist tendencies in its common arrangements for agriculture. But there was no escaping the fact that the world's largest importer of non-tropical agricultural products was indicating its willingness to embrace the common farm policy with the Six, and but small chance that its more trade-friendly deficiency payments system of supporting its farmers' incomes could survive. Moreover, in so far as the United Kingdom was successful in its declared intention to protect the agricultural trade interests of its Commonwealth suppliers and EFTA associates, this could only widen the discrimination against third-country exporters. It was not until 1963 that General de Gaulle vetoed Britain's membership of the Community, so the prospect of the further agricultural trade dislocations that would attend Britain's joining the Community still hung over the Dillon Round in its closing stages.

A third development with negative implications for agricultural trade was the treatment of agriculture in the two other regional preference areas that were formed in the period under review. Seven non-EEC countries formed the European Free Trade Association (EFTA) in 1960, and a number of countries in South America joined together in the Latin American Free Trade Area (LAFTA) in the same year. Neither agreement conformed fully with the intent of the GATT Article XXIV, most notably the EFTA which excluded the agriculture sector in apparent violation of the requirement that barriers be removed on 'substantially all the trade' between the member countries. Both agreements were examined by the CONTRACTING PARTIES (the EFTA more intensively than the LAFTA since the signatories were already taking a relaxed view of regional trading arrangements amongst developing countries). In the case of the EFTA, after a Working Party failed to reach agreement on whether Article XXIV meant 'substantially all the trade' or 'trade in substantially all products' (BISD 9S/85), the CONTRACTING PARTIES allowed the EFTA to go into effect.

5 CONCLUSION

The Dillon Round of the GATT negotiations came to an end in March 1962, with the conditions of trade in agriculture even worse than they had been when the Haberler Committee reported in the autumn of 1958. Agricultural products had been largely unaffected by the minuscule tariff reductions achieved in the second phase. In the first phase, the EEC had cleared the way for the introduction of a common agricultural policy by denouncing its GATT obligations with respect to previously bound tariffs on important farm products (themselves only the least important of the array of policy instruments used to protect and support Europe's farmers). The Community had refused to enter into any equivalent commitments on maximum levies and minimum access. During the course of the Round, the rest of the world was shown the framework of the common policy that was to be introduced and expected that the exceptionally trade-distorting set of policy instruments chosen would be used to defend prices determined by the income needs of the operators of Europe's small farms. If successful, the United Kingdom's application for membership would place that country behind the import levy wall. The 'standstill agreement' the Community had signed covered only a few products and its worth was not immediately evident. Nor did the fact that the Community had acknowledged that the United States (and other exporters of hard wheat) had 'unsatisfied negotiating rights' seem to be a factor with any weight in the Community's thinking about its future agricultural support arrangements. The only suggestion from the Six as to how the adverse effects on the conditions of world agricultural trade of such policies might be attenuated was M. Baumgartner's ill-defined proposal for the creation of a highly dirigiste regime, which seemed to take trade in agriculture out of the GATT completely.

The United States permitted the Dillon Round to conclude because of the strategic importance it attached to the Community's forming its customs union, embarking on its first common economic policy, and enlarging its membership. But in so doing the United States had failed to reduce the potential for trade distortion, had allowed the Community to default in its tariff obligations under the GATT, and had acquiesced while Europe embarked on the introduction of the common agricultural policy.

The tide of events in the world polity and economy – the completion of European recovery, the formation of the Community and its prospective enlargement, the emergence of Japan as an economic giant, the faltering of the economy of the United States and its diminishing ability to act as the hegemon in shaping international economic and security relations – was too much to be dealt with by a GATT round limited to rebalancing tariffs and reducing them by the small degree permitted by the authority vested in the United States Administration by Congress. It needed a bolder initiative in commercial diplomacy if the United States were to come to satisfactory terms with the emergent EEC and to ensure that it evolved in ways that were internationally constructive. Accordingly, planning for a more ambitious round began shortly after the Kennedy Administration took office, and even before the Dillon Round was completed. From the outset, it was clear that agricultural trade reform was to be the centrepiece of what would come to be called the Kennedy Round.

4 The Kennedy Round[1]

1 THE SETTING

President J.F. Kennedy was inaugurated in January 1961. It was an important moment in world affairs. European economic recovery was completed. The European economies were individually strong. The Six had embarked upon the formation of an economic union, and Britain and other EFTA countries were expected to join soon. The economic union was developing with unexpected speed. 'Europe' was beginning to take on a distinctive identity, and some voices in Europe were saying that the influence of the United States in political, economic, and security matters should be diminished.

The United States was the undisputed champion of the free world, but its ability to shape world affairs was undermined by its faltering economic performance. Its gold reserves were now less than the sum of foreign-held dollar obligations, and its balance of payments was in deficit. It needed to form a new alliance with Europe in which the burdens of containing political and military adversaries, managing the world's monetary and trade arrangements, and promoting third-world development could be shared.

A balance of payments surplus had to be earned if US security obligations, private foreign investment, and development cooperation were to be sustained. Agriculture was thought to be a major potential contributor to that goal. The sector already provided a merchandise surplus, and the overall comparative advantage of the United States in agriculture was firmly based on the industry's productivity. Moreover, exports were an increasingly important contributor to farm income. The instrumentation of US farm policy had been changed for the basic export crops by replacing high loan rates and export subsidies with deficiency payments. Thus United States trade and agricultural policies were predicated on success in expanding exports of farm and food products.

To the new Administration, a major trade initiative focused on the North Atlantic countries offered multiple benefits. In economic terms, reducing tariffs would minimize the threat of trade diversion and the redirection of international investment inherent in the formation (and

52

enlargement) of the EEC, and it would help restore balance to the country's external accounts. In political terms, a successful trade negotiation would deny schism amongst the industrialized democracies and lay the foundations of an 'Atlantic Alliance' between the two great common markets of the free world. Thus the Kennedy Round was the commercial vector of America's attempts to forge a new political, military, and economic partnership with the emergent Europe. Its objectives were to maintain access to a uniting Europe's markets and to place the 'Atlantic Alliance' on a sound economic base.

Agricultural trade was central from the beginning for a variety of reasons. Expanding United States farm exports would contribute to the necessary improvement in the balance of payments and sustain farm incomes. Beyond that, it would indicate whether the trend in the world trading system was to be towards multilateral liberalism or regional protectionism. Access to the all-important European market for farm exports could only be improved in the context of comprehensive trade negotiations, because improved access for US farm exports had to be secured by offering improved opportunities in the American market for Europe's manufactures. Congressional support for a bold initiative in trade liberalization would only be forthcoming if the negotiations held the promise of rolling back the protectionism inherent in the EEC's common agricultural policy.

Only the United States had the ability to initiate a new GATT round, and it was motivated to do so by its strategic, economic, and agricultural objectives. Also, because of the practice of reciprocity in GATT negotiations, the United States was the only country able to make access offers large enough to persuade the Community to lower its common external tariff. But other countries welcomed the prospect of a new GATT Round too. At the time, the Community was anxious to demonstrate its 'outward orientation'. Within the Community, West Germany was making participation in external trade talks a precondition for progress on forging common internal policies. The EFTA countries saw multilateral tariff reduction as a means of minimizing the dangers of trade diversion in Europe. The place of Japan in the international trading system needed to be addressed in a multilateral framework. The smaller agricultural exporters were eager to obtain improved access for their agricultural products to markets not only in Europe, but also in North America and Japan. And the time was at hand for the developed countries to deliver on the promises they had been making since the Haberler Committee had reported in 1958 about their willingness to improve the trade conditions facing the developing countries. Thus, the times were propitious for a major trade negotiation, and most countries had positive incentives to see it launched.

However, while acknowledging the multiple interest, the Kennedy Round was in practice a negotiation between the United States and the European Economic Community. Japan was not yet a superpower and was still quiescent in international economic affairs. The EFTA countries were primarily concerned with intra-European trade in manufactures. The Commonwealth exporters of non-tropical agricultural products were individually small and not organized to negotiate as a group. The developing countries had little to offer and had highly differentiated demands. The Kennedy Round was destined to be a gladiatorial contest between the established but faltering world leader and the emerging entity in Europe.

United States farm leaders and Congress insisted that the negotiations must yield a significant measure of trade liberalization for agriculture. Political leaders and top-ranking officials committed themselves to delivering on this promise. At home and abroad, they insisted that, for the first time since its inception, agriculture must be brought fully within the rules and disciplines of the GATT and that there would be no agreement in the pending negotiations to liberalize industrial trade unless trade in agriculture was also freed. As the negotiations began, President Johnson assured the members of his Public Advisory Committee on Trade Negotiations that 'The United States will enter into no ultimate agreement unless progress is registered towards trade liberalization on the products of our farms as well as our factories' (US Department of State, 1964). This stance was maintained despite 'an early and deep pessimism about the eventual agricultural outcome ...' among senior officials.[2] That this pessimism was well founded was soon to be confirmed.

2 PREPARATIONS AND PRENEGOTIATIONS

The Kennedy Round of comprehensive trade negotiations officially opened in Geneva in a three-day session in May 1964 and was concluded there on 30 June 1967. The Round's origins can be traced back to 1958, and 'negotiations about negotiations' can be said to have begun in November 1961.

By the time GATT ministers met in late November 1961, even before the Dillon Round had reached its disappointing conclusion, it was apparent that a major new trade initiative was needed. The 1961 GATT ministerial meeting 'culminated the program for the expansion of international trade inaugurated at [the] ministerial meeting in November 1958' (Preeg, 1970, p. 41) at which the Haberler Report had been considered. In their final communiqué, the GATT ministers set in motion the preparations for a new round. The ministers foresaw that such a new negotiating round would need to use 'some form of linear tariff reductions' to liberalize trade in manufactures. They reaffirmed the need for 'progressive reduction and elimination of barriers to the exports of less-developed countries'. And, having noted the degree and extent of agricultural protectionism and the widespread use of non-tariff devices, the ministers invited the CONTRACTING PARTIES to work out negotiating procedures 'for the creation of acceptable conditions of access to world markets for agricultural commodities' (BISD 10S/25–8).

In preparation, in January 1962, the Kennedy Administration had submitted the Trade Expansion Act to Congress, and this had signed it into law in June that year. The new law sought an escape from the constraints in the Trade Expansion Act of 1960 that had prevented the United States from responding fully to the offer the EEC had made in the Dillon Round to cut its common external tariffs by 20 per cent. The new Trade Expansion Act gave the Executive authority to cut all tariffs by 50 per cent and to zero on products on which the United States and the EEC accounted for 80 per cent or more of world trade.[3] It also removed the 'no-injury-to-domestic-industries' constraint by replacing the 'peril-point' procedure with a programme of adjustment assistance. In promoting the Act before Congress and in the country, Administration spokespersons made clear that a major aim was to reduce European agricultural protection. In passing the Act, Congress made clear its intent in the agricultural area by authorizing the President to adopt retaliatory measures against countries that used 'variable import fees' that penalized US agricultural exports.

GATT ministers met in May 1963 to finalize arrangements for the Round. They agreed negotiations should begin on 4 May 1964. They created a Trade Negotiations Committee to elaborate a detailed trade negotiating plan and to handle the overall negotiations. Subcommittees were formed to deal with tariffs, non-tariff measures, the participation of the developing countries, and the negotiations on agriculture.

The overall objective for agriculture was contained in the injunction that 'the trade negotiations shall provide for acceptable conditions of access to world markets for agricultural products' (BISD 12S/47–9). The Trade Negotiations Committee was charged with the task of establishing 'the rules to govern, and the methods to be employed in, the creation of acceptable conditions of access to world markets for agricultural products in furtherance of a significant development and expansion of world trade in such products.'

The communiqué went on, 'Since cereals and meats are among the commodities for which general arrangements may be required, the Special Groups on Cereals and Meats shall convene at early dates to negotiate appropriate arrangements. For similar reasons a special group on dairy products shall also be established.'

These groups were to report to the Committee on Agriculture.[4] Christian Herter, the US Special Trade Representative, stated that while improving the conditions of access was the United States' primary goal, the government was 'prepared to negotiate, within the context of [such] agreements, its production, price, export and import policies on a reciprocal basis' (GATT, 1963).

The 'negotiations about negotiations' however were not started immediately after the May 1963 GATT ministerial meeting. France withheld its approval of the negotiating proposal to be tabled by the Community in Geneva until it had shaped the CAP to its vision in Brussels. The key elements of France's position were that it would accept internal and external free trade in manufactures only in return for a common agricultural policy having six key features: that Community agriculture would be protected; that Community farmers would have preference in selling into the European market; that grain prices should be unified and set at a level acceptable to French growers (but not so high as to stimulate production elsewhere in the Community); that there should be common financing of the CAP; that no access guarantees should be given to third-country exporters; and, finally, that agreements on these matters among the Six should not be unravelled in the upcoming international trade negotiations (Coombes, 1970, pp. 167–216). Throughout 1963, the French forced a series of crises in Brussels on financing the CAP, common grain pricing, and regimes for other products. In between, Britain's application for membership in the Community was vetoed *inter alia* because the British might attempt to shape the CAP to an alternative vision. Progress towards a Community negotiating position on agriculture in the Kennedy Round was the hostage in each crisis.

It was not until December 1963 that the Commission was able to break the internal and external log jams simultaneously with proposals that came to be known as Mansholt Plans I and II (Mansholt was EEC Commissioner for Agriculture at that time). Mansholt I was a proposal to establish unified cereals prices in a single step, with compensatory payments being made for a three-year period to farmers in the countries where grain prices would fall. Mansholt II was a plan for a joint agreement on common financing and the mandate for the GATT negotiations. The central feature of the latter was acceptance of the *montant de soutien* (degree of support) as the central element in the Community's negotiating offer (see below). The package was agreed, and the Community formally submitted its proposal on the *montant de soutien* to the GATT Committee on Agriculture (which was considering the manner in which agriculture should be negotiated) in February 1964 (GATT, 1964a). It was discussed in conceptual and technical terms in the Cereals Group during the spring and summer of 1964, but before negotiations proper could get underway, another transatlantic firestorm had to be extinguished.

3 THE CHICKEN WAR

The notorious 'chicken war' occurred in the period between the preliminary decision to embark on major negotiations (November 1961) and their formal opening (May 1964). The application of the Community's sluicegate and levy system to poultry from July 1962 had an immediate and adverse effect on the United States' recently established but rapidly growing exports of frozen chickens to West Germany. The effect of the levy was to raise import charges from 15 per cent to an ad valorem equivalent of 40 per cent by December 1962, and subsequently even higher. As a result, US chicken exports to West Germany plummeted. The application of the 'feed-grain compensatory levy' was legal in so far as the previous 15 per cent duty on poultry entering West Germany had been unbound in the Dillon Round, but the increase in border charges was in clear violation of the 'standstill' agreement reached with the Community at the conclusion of the Round in March, only weeks earlier. Accordingly, the United States mounted an intensive bilateral effort to obtain modification of the levy. When this failed the US took the matter to the GATT on the basis of the agreement it had with the EEC. The Community did not deny the validity of the United States' complaint. However, there was a dispute about the damage done to US exports and hence on the amount of compensation which the United States was entitled to receive by withdrawing 'substantially equivalent' concessions. This amount was eventually established by an impartial GATT advisory panel, and in December 1963 the US increased duties on light panel trucks (targeted at the

German Volkswagen), brandy (targeted at France), potato starch, and dextrine.

The importance of the dispute went far beyond the few millions of dollars of poultry exports involved. The United States was testing the integrity and value of the 'standstill' agreement and 'unsatisfied negotiating rights' which the Community had agreed that it held at the end of the Dillon Round, not just on poultry but also on grains, the exports of which were many times greater in value. It was making the point that it still expected to be compensated for the losses it had experienced by the CAP's disturbing its access to the individual markets of the Community's member states. Secondly, the uncompromising stance the United States took on poultry was intended to show the Community (and domestic constituents) that it was going to take a hard line on agriculture in the multilateral negotiations that were about to open. Hence, major issues surrounded what appeared to be a minor trade dispute. 'The heat generated by frozen chickens' (Evans, 1971, p. 174) caused a great deal of rancour in the lead-up to the Kennedy Round negotiations.

4 THE ROUND DELAYED

The year and a half from the GATT ministers' agreeing in principle that a new round should be held (November 1962) to the opening of the Kennedy Round proper (May 1964) contained much to discourage the United States and the smaller and developing exporters of non-tropical agricultural products. Since December 1962, the CAP had emerged with all the features that threatened to reduce access to the European market and to generate subsidized competition with the Community's excess supplies in third markets: high support prices, variable import levies, community preference, and export subsidies for intervention stocks. The United Kingdom and other EFTA countries had indicated their willingness to adopt the CAP if they were permitted to join the Community. The chicken war – the first real encounter with the CAP – had demonstrated the large, abrupt and cumulative damage that the CAP could wreak on trade. The Community had refused to give the exporters interim guarantees of access to its market pending the outcome of the GATT negotiations. The 'comprehensive commodity arrangements' suggested by Community spokespersons and being explored in the Cereals Group were not appealing to those who wanted agricultural trade liberalized rather than managed. And, for reasons that are discussed below, the Community's proposal that binding the degree of support provided to farmers by national agricultural policies be the central feature of an accord on agricultural trade reform did not appear to offer exporters the improved access they were looking to the Kennedy Round to provide.

The formal opening of the Kennedy Round was also discouraging, for it was apparent that the preparatory work of the Trade Negotiations Committee and the subcommittees had not resulted in agreement on the rules and timetables for the negotiations, or on the modalities to be employed in finding solutions in each of the four areas of the negotiations – tariffs, non-tariff measures, developing-country trade, and agriculture. It was to be six months before negotiations could actually begin, and then only on industrial tariffs. Negotiations on agriculture were to be subject to an even longer delay. The position of the United States was that negotiations on agriculture and manufactures should move together. The EEC, however, was not yet ready to make offers in either sector, since France and the Commission were linking the Community's industrial tariff offer to agreement being reached on the common prices that were to prevail for cereals in the EEC. In November 1964 the United States reluctantly, and against the wishes of the US Department of Agriculture, allowed negotiations on industrial tariffs to begin without comprehensive agricultural offers having been made.

In January 1965, Mr Eric Wyndham White, the Secretary-General of the GATT, attempted to break the stalemate in agriculture by proposing that specific offers for each commodity be made – for cereals by 1 April 1965, and for meats, dairy, and other products by September 16 (GATT, 1965). Offers on cereals were tabled by all the major countries, but not the EEC, by mid-May. Talks aimed at clarifying the offers began in June. However, substantive negotiations aimed at bridging the gaps between them could not be joined, because on 30 June 1965 the French precipitated another major crisis in the Community by beginning a seven-month-long boycott of all Community business. The 'empty chair' tactic of France in Brussels stymied progress in Geneva. Though technical discussions in Geneva did continue, active negotiations ground to a halt. It was 30 January 1966 before the Six reached an agreement – the Luxembourg Compromise – and were able to move on to further internal agreements on financing the CAP, and (in July 1966) to the details of the market organizations for dairy, beef, sugar, and fruits and vegetables. With internal agreements between the Six finally in place, the Community completed its GATT offer on agriculture on 5 August 1966. More than two years after the official opening of the Kennedy Round, 'a virtually complete set of offers was on the table' (Preeg, 1970, p. 125), and the substantive phase of the negotiations on agriculture could begin.

5   SUBSTANTIVE OFFERS

Negotiations on the agricultural offers began in September 1966 and ran until the final hours of 30 June 1967. However, 'discussions' on agricultural trade had begun way back in February 1962 when the Cereals Group first met to consider alternative ways in which the world trading system for grains could be improved. On the table at that time were two proposals which represented the opening (and polar) positions of the major protagonists. The Community had suggested the essentials of the Baumgartner Plan as the basic approach to reforming world trade in agriculture.[5] The counter opening proposal of the United States was that, in principle, all non-tariff barriers to agricultural trade should be converted to tariffs, which would then be reduced and bound.

For public consumption at least, the fundamental position of the United States was simple. The goal was to return to the original promise of GATT – the promise of a market-oriented agricultural trading world. GATT rules should apply equally to agricultural and industrial products. In principle, tariffs were the only legitimate trade barrier. These should be applied in a nondiscriminatory manner and progressively negotiated away so that the location of agricultural production could be determined by comparative advantage working through market mechanisms. The CAP was highly protectionist. Since the purpose of the negotiations was to expand efficient trade, the protection afforded by the CAP should be reduced. The variable import levy was an illegitimate instrument since it gave absolute protection, was variable, forced third countries into the position of residual suppliers, and, unlike a tariff, denied them expanded trade opportunities as their productivity increased. Levies should therefore be replaced by tariffs, cut by the 50 per cent proposed for manufactures, and then bound against increase. Where measures other than tariffs were used and these could not be replaced, then the aim of the negotiations should be to secure trade liberalization equivalent to a 50 per cent cut in protection.

This position, which has been described as 'at best visionary' (Evans, 1971, p. 203), was quickly abandoned, and by the time the Kennedy Round opened in May 1964 there was a general understanding that trade in farm products should be handled in three ways. Those products that were subject to duty-free bindings (such as soya beans) should be left as they were. Those for which protection was afforded by tariffs should be subject to tariff reductions and bindings, with the tariff cuts to approach, as closely as possible, the 50 per cent linear reduction that was the working hypothesis guiding the negotiations on industrial products. Thirdly, there was an understanding that the major bulk commodities – more especially, those that would be subject to the levy and export 'restitutions' regime of Europe's CAP – would be handled by some form of 'comprehensive commodity arrangements'.

The Baumgartner Plan

The Cereals Group had begun consideration of one vision of a world grains arrangement in February 1962. The previous November, at the sixteenth Session, in a discussion of the Third Report of GATT's Committee II, the Minister of Finance and Economic Affairs for France, M. Baumgartner, having declared categorically that the rules of free competition could not be applied to agricultural products, went on to give a first glimpse of a generic scheme for organizing world trade in temperate zone agricultural products on the basis of comprehensive commodity agreements.[6] Though never spelled out in detail either by M. Baumgartner in the GATT or by other French spokespersons who raised the matter in other fora, the gist of the idea was that world agricultural trade should be organized according to a pattern that combined elements of the European Community's proposed Common Agricultural Policy (CAP) and the surplus disposal programmes operated by the United States under PL480. There were also elements that suggested globalization of the pricing and guaranteed access arrangements for tropical products that metropolitan France operated with its overseas territories. Exporters would offer their products at agreed 'reference' prices, while rich-country importers would impose levies to bring the landed price of imported products up to their internal support levels. Part of the increased revenue of the rich exporters and the levy income of the developed importers would be used to dispose of excess supplies as food aid to developing countries.

Such a world grains arrangement would have to satisfy multiple objectives including balancing supply and demand, limiting national subsidies, stabilizing international prices within an agreed range, improving access to import markets for efficient exporters, sharing the costs of food aid, and coordinating the noncommercial disposal of surpluses. There were, of course, early and deep differences between countries on the weight to be given to each objective and the means for its attainment. Above all else, the exporters wanted assurances about their future access to import markets, and wanted these to be provided both by 'access guarantees' and by international disciplines on the price supports in importing countries which, in the absence of such constraints, would cause output expansion and import substitution. It was never apparent how M. Baumgartner's scheme – the Baumgartner Plan – could contribute to these exporter objectives. Discussion of the plan continued for some time in the Cereals Group, but it was eventually joined and overtaken by discussion of a more tangible proposal on agriculture from the EEC, that of binding the degree of support, the *montant de soutien* (MDS).

Montant de Soutien

The EEC's posture in the negotiations was liberal in all areas except agriculture. Its common external tariffs on manufactures were already lower than those of most other countries, and it was willing to negotiate on the basis of a 50 per cent linear reduction. But the Community took the view that agriculture really was 'different' by reason of the unique characteristics of both the supply and demand sides of the market for farm products, and the universal presence of agricultural support policies. In all countries, frontier measures were but one of the many interventions determining the degree of protection given to farmers by national agricultural programmes. Hence, in the EEC's view, the appropriate negotiating procedure was first to measure the extent of the protection given to producers of each commodity in each country, and then to bind the degrees of protection established against increase in negotiations. Binding the MDS would be equivalent to binding a tariff.[7] Having offered to bind the degree of support it provided its farmers, the Community resisted the continuing demands of the US and other exporters that it offer quantitative minimum access guarantees for agricultural imports.

The EEC proposed that each country should bind the degree of support it provided to its producers in relation to a set of world reference prices. The *montant de soutien* would be a common measure of the degree of support provided to producers in all countries by all policy instruments that affected production and trade, including market price supports and direct payments for products, and input and export subsidies. The reference prices from which support would be measured could be an average of actual market prices over an agreed period, or the minimum (or 'floor') prices established in any intergovernmental commodity agreements that might be negotiated. Importers, such as the EEC, would charge a fixed levy equal to the difference between the reference price and the internal support price. However, they would be free to impose a supplementary levy if exporters allowed their offer prices to fall below the reference levels.

In the EEC's view, the negotiated maximum degrees of support for national producers could be but one element in a comprehensive grains agreement. As noted, the reference prices from which support would be measured could be the floor prices for each grain in such an agreement. If the exporters wanted prices stabilized at higher levels than the prices then prevailing in world markets, so be it. The reference prices in the world grains arrangement were negotiable. 'Respecting' the floor prices would put an end to export subsidy competition and compensate exporters through higher prices for any reduction in the volume of sales they experienced. The exporters were anxious to have the rich importers share the costs of food aid to developing countries. The Community suggested that there could be a link between surplus disposal obligations and price stabilization objectives.

The central feature of the Community's proposal – the MDS concept – had the virtue of acknowledging that trade problems in agriculture arose from national farm price and income support programmes and that the degree of distortion was a function of support levels. It had the merit of cutting across the maze of measures that governments used to assist their farmers and directing attention to the trade distortions that arose from the degree of subsidization and protection. It promised to limit the adverse effects on world markets of unfettered national farm income support programmes, while leaving governments with some autonomy in the instruments they employed. Furthermore, binding in an international agreement the level of income support provided to farmers through commodity prices implied that future income increases for farmers would have to come from direct income payments, productivity gains, and structural reforms.

In the judgement of the EEC, binding the degree of support offered by the CAP represented a real concession, and one that had been difficult to wring from the member countries. To the EEC's dismay, the United States reacted negatively to the proposal when it was first aired in 1962, and again when it was put forward formally in 1964, and the US maintained its antipathy when the proposal was repeated in May 1965 and July 1966. With hindsight it seems regrettable that the exporters did not seize the offer of the Community to enter into a binding commitment to place a ceiling on the support the CAP would provide to Europe's farmers. But, at the time, there seemed to be good reasons for the US' stance. In principle at least, the exporters had gone into the Kennedy Round hoping to have the applicability of existing GATT rules affirmed. The *montant de soutien* seemed to require a special code for agriculture – something which had often been proposed since the GATT's inception and steadfastly resisted. The concept had no substance in the early stages of discussion since the CAP's support prices had not been established. When they were, and when the Commission was pressed to turn the concept of the MDS into a negotiable proposal, it became clear that the Community could only contemplate binding or 'consolidating' the very high degree of support that CAP prices provided (GATT, 1964a). As Schnittker put it

> The European Community, having determined to increase the differential of internal prices over world prices prior to proposing the montant de soutien method of negotiation, was ready for all countries to agree that the level of protection was high enough. (Schnittker, 1970a, p. 442)

This was not what the exporters had gone to Geneva to secure. They wanted the level of protection reduced so that export opportunities would expand, and the Community was not prepared to meet this acid test. Moreover, there was every prospect that at the high levels of support provided by the CAP, Europe's output could increase and its net import requirements fall. The Community was not contemplating introducing supply controls, or even limiting the quantum of output eligible for its price guarantees. Since, under the EEC's proposal, the exporters would be offered no quantitative access guarantees, there would be nothing to prevent European grain growers, especially French producers, taking over the European cereals market. Furthermore, since the Community proposed that the reference prices should be renegotiated every three years, it was possible that the support bindings would also be bound only for three years, unlike tariffs which were conceptually bound indefinitely. Acceptance of the proposal would have implied endorsement of the unpopular variable import levy, the legality of which under the GATT was unclear and, as yet, untested. The responsibility for sustaining the world reference prices – by adjusting output, stocks or noncommercial disposals – would have fallen entirely on the exporters. At the time, it also seemed that there would be immense technical problems in measuring the degree of support provided by all programmes and policy instruments for the multiple forms and grades in which farm products were produced and traded.[8] Later, the US' and other exporters' hostility deepened when the Community began talking about the *montant de soutien* as a generalizable solution to the problems of agricultural trade. Extending the concept to other commodities – including those in which protection was provided by fixed tariffs, some bound at low or zero rates – would have worsened access and eliminated price as a mechanism for determining production and trade. Finally, the Community's offer was conditional on other countries' binding their present levels of price support, a condition not formally rejected by the exporters, but for which they were ill prepared, and for which the US delegation had no political or legislative mandate.

As the months drifted by, the Community's suggestion that support margins be bound seemed to become more ambiguous, and even more a concept rather than a concrete offer on which substantive negotiations could begin. Though the MDS stayed on the table in the Cereals Group it gradually receded as an option.

Binding Self-Sufficiency

The Community eventually responded directly to the grain exporters' earlier proposal that they be given assured access to the European market. The EEC's Council of Agricultural Ministers agreed to a proposal of the Commission that the Community limit its grain output to an agreed percentage of Community consumption. In July 1966 this translated into a proposal made in the Cereals Group that countries fix their 'self-sufficiency ratios' for grains at levels to be agreed. Production in excess of agreed market share would be kept off commercial markets by 'slopping over' into food aid supplies. This would protect the floor prices in the proposed world grains agreement, maintain access for commercial exporters, and force the countries that allowed their grain output to exceed agreed limits to bear the costs of disposing of the surplus as food aid.

In principle, this was a development that could be welcomed by the exporters, for it would provide them the access guarantees they had sought as early as the Dillon Round, and had been requesting throughout the prenegotiations and negotiations in the Cereals Group, and it was a key element of a proposal that the US and smaller grain exporters had themselves made in the Cereals Group in May 1965 for an international grains arrangement.[9] Moreover, it would force importers to discipline their output or share in the cost of providing food aid to developing countries. Unfortunately, when the time came to table concrete numbers, the EEC and the United Kingdom proposed self-sufficiency ratios for themselves of 90 and 75 per cent respectively (Curtis and Vastine, 1971). These figures were unacceptable to the exporters, for they were higher than the current degrees of self-sufficiency (86 and 60 per cent), and even higher than the levels being projected by the EEC itself at the time for its level of self-sufficiency in 1970. Additionally, the Community proposed that the maximum self-sufficiency rates would be bound only for three years, and then only when the Secretariat supervising the operation of the International Grains Arrangement declared that a surplus situation existed. At the time, this seemed a meaningless offer in trade terms, for the exporters were being asked to agree to a deterioration of their position in the West European grain market. That was not why they were in Geneva.

6 RESOLUTION

Discussion of the Community's proposal on self-sufficiency ratios and on various ideas about the content of a comprehensive grains arrangement dragged on until May 1967. By then it was judged that the Community was not willing to make commitments on levels of support and access that were meaningful in trade negotiating terms to the United States and to the smaller exporters. Accordingly, with the United States delegation now being led by a new 'results-oriented' Special Representative on Trade Negotiations, Ambassador William Roth, the final weeks of the Kennedy Round negotiations on grains were spent on the other matters that were to be included in a comprehensive grains agreement – wheat prices and food aid. The result was the initialling, on 30 June 1967, of a *Memorandum of Agreement on Basic Elements for the Negotiation of a World Grains Arrangement* (GATT, 1967). The International Grains Arrangement (IGA) was negotiated in July and August 1967, in Rome, under the auspices of the FAO. The IGA had two components: the Wheat Trade Convention and the Food Aid Convention.[10]

The Wheat Trade Convention (WTC) was melded into the administrative and institutional mechanisms of the existing International Wheat Agreement. Its key provisions were an increase in the floor and ceiling prices for wheat; the designation of fixed differentials for 14 specific grades of wheat; the replacement of Thunder Bay, Canada by US Gulf ports as the f.o.b. price basing point; a 'best endeavours' commitment of the importers to buy from, and the exporters to supply wheat to, signatories at prices within the range; and the establishment of a Price Review Committee to oversee and make recommendations for changes in the operation of the Convention. The WTC had no provisions dealing with access, market sharing, supply management, price support disciplines or precise obligations on actions to be taken in defence of the price range. The USSR – which at the time was still an important wheat exporter – was not a signatory.

The Food Aid Convention (FAC) committed the signatories to supply annually 4.5 million metric tons of wheat, or its equivalent, as aid to developing countries on the basis of agreed shares. The United States' share was 42 per cent and that of the other net exporters was 16.5 per cent. The importers together provided 1.9 million metric tons or 42.5 per cent, with the EEC's share being 1 million metric tons (23 per cent) and Britain's and Japan's 25 000 metric tons (5 per cent) each. The FAC had the merits of having the developed grain importing countries share with the exporters the cost of providing food aid to developing countries. At the time, it was thought that food aid would be needed on a vast and growing scale for disaster relief and development assistance in the developing world.[11] The multilateral commitment should also have reduced the size of stocks overhanging the world market and, so long as the importers did not simply increase production, provided space for another 1.9 million metric tons of commercial sales to the developed importing countries. However, the total was far below the 10 million metric tons that had been set as the goal at the beginning of the negotiations, and, of course, there was no link between grains donations and access opportunities via self-sufficiency ratios, nor any link between food aid and the defence of the floor price for wheat set in the Wheat Trade Convention. In the end, Japan insisted that it be permitted to meet its commitment with other aid (for example, cash or fertilizers) rather than grains.

The abrupt decision of the United States in the closing stages of the negotiations to limit the discussion on grains to securing a higher floor price for wheat and a multilateral commitment on food aid was based on a number of considerations. There was thought to be no trade advantage for wheat if the US had agreed to the Community's binding the level of support at the existing high level, and extending the *montant de soutien* to such products as soya beans, tobacco, and fruits and vegetables, as the EEC proposed, would have been a regression from the tariff bindings that the US held. Beyond that, 'acceptance of negotiations on the basis of the MDS. . . would concede acceptance of the principle of the variable levy system, and even tacit endorsement of it.'[12] Market sharing arrangements would have been unacceptable to Congress at the self-sufficiency levels proposed. This was particularly the case for feed grains for which the market prospects were thought to be good, and where the American Farm Bureau Federation and the grain trade focused their opposition to market sharing. Further, the EEC's suggestion that the US and other exporters should bind the degree of support and apply market sharing concepts to their own support programmes for wheat – and potentially more broadly – was politically unacceptable and economically harmful to the sectors with good growth potential such as feed grains, oil-seeds, and meats. The dangers in continuing discussions along these lines were averted. Furthermore, the United States considered that it still had the 'unsatisfied negotiating rights' salvaged from the Dillon Round which it could exercise at some future date (Evans, 1971, p. 271).

The negotiations on the other major commodities – dairy products and beef – fell even further short of attaining the exporters' goals than the negotiations on grains. New Zealand wanted an international arrangement on dairy products and was attracted by the proposal to fix minimum world selling prices contained in M. Baumgartner's plan (Curtis and Vastine, 1971, pp. 41–50). At a minimum, New Zealand and other exporters of dairy products wanted limits placed on export subsidies and the diversion of surpluses to a multilateral food aid programme. The United States was interested in moderating dairy protection in the Community and in preserving access for exporters to the important United Kingdom market, largely because this would relieve the pressure for the US to open its own tightly controlled borders to larger dairy product imports. The United States quickly lost interest when, to its chagrin, the exporters concentrated their fire in the Dairy Group on the US' Section 22 import quotas for manufactured dairy products. With the EEC unwilling to give assurances about the trade effects of its price support arrangements for milk, the United Kingdom planning to renew its application for membership of the Community, and neither Japan nor the United States willing to open its highly protected markets to dairy product imports, nothing could be accomplished in the Dairy Group. The desultory discussion of the possibility of a commodity arrangement for the world dairy sector was abandoned in early 1967.

The United States would also have welcomed some improved access for Argentinian beef shipments to Europe, and assured and improved access for shipments from Australia and New Zealand to Britain and Japan, for this would have relieved the pressure of Oceanic beef imports on its own market. However, the US was not interested in opening its own borders to beef exporters. Indeed, in 1964 even as the Kennedy Round began, the US, in response to a coincident world beef surplus and domestic cyclical over-supply, introduced the Meat Import Law which placed meat under quantitative import restrictions. This might have been challenged as being in violation of the GATT's Article XI, but the US had its 1955 waiver from the disciplines of that article. With the United States closing its market to imports it was not to be expected that other countries would be willing to open theirs. The discussions in the Meat Group on access and minimum world selling prices came to naught, and the work of this group was also suspended early in 1967.

The work of the Dairy and Meat Groups might have been more successful had progress in the all-important Cereals Group been possible. As it was, in the last few weeks of the negotiations, efforts were concentrated on bilateral negotiations between the major importers and exporters of dairy products and meats on a request and offer basis. This led to some marginal adjustments to import tariffs and quotas on dairy and meat products.

7 OVERVIEW

Overall, the Kennedy Round was a successful attempt to further liberalize and discipline world trade. A new technique of reducing tariffs, linear cuts, was used for the first time. Duties on thousands of tariff line items were reduced by 50 per cent, and an average tariff reduction of 35 per cent was achieved on $40 billion of trade. Work on non-tariff measures (NTM) affecting trade was begun. A new anti-dumping code was written, and the foundations were laid for the major assault on NTM that was mounted in the Tokyo Round which was to follow in the next decade.

The Kennedy Round came short in the areas in which the GATT had had difficulties throughout the twenty years of its existence. Little was done about the particular trade problems of the developing countries; indeed, the conditions for trade in textiles and apparel regressed. The accomplishments in agriculture were modest indeed. At the end of the negotiations, the tangible results in agriculture were some tariff reductions on farm and food products and the International Grains Arrangement. Tariff cuts of 50 per cent or better were made to $1.6 billion of agricultural trade, and the average reduction achieved was 22 per cent.[13] The Wheat Trade Convention was the major result that

US agricultural negotiators took home from Geneva. Alas, the chief benefits it promised – higher prices for wheat exports – failed to materialize. Within 12 months of its coming into force its floor prices were breached. In the absence of provisions on joint supply management and market sharing, the pricing provisions never took effect. The Food Aid Convention survived, but it had lost much of its significance with the failure to link it to negotiated maximum levels of self-sufficiency in the major import markets. This was a meagre return indeed for so much effort.

The Round had failed to attain its avowed objectives in agriculture – reducing the trade distortions caused by protectionist national agricultural policies and providing improved and assured access to import markets for efficient exporters. By the end of the Kennedy Round, the European Economic Community's highly protectionist common agricultural policy had been created and put in place. North America, Japan and the EFTA countries had not opened their markets for the farm products they imported. And no government's autonomy in the design, instrumentation, and operation of its national agricultural policies had been constrained. Though agriculture had occupied the attention of the political leaders and senior officials of the world's great trading powers for four years, the first serious attempt to bring agricultural products within a liberal and disciplined multilateral trading system had failed. On the central issues, nothing had been accomplished.

The focus of the agricultural negotiations was the attempt by the United States to shape the CAP in ways that would preserve the market in Europe for US exports of grains. Japan and the United Kingdom were not pressed on grains. And when Australia, Argentina, and New Zealand pressed these countries and the United States for improved access for their exports of beef and dairy products, US negotiators drew back for they had neither the inclination nor the authority to open the US market for import commodities.

Although the United States and the other exporters had pervasive and deep-rooted objections to Europe's common policy for grains, their initial negotiating approach was pragmatic in so far as it emphasized access guarantees. Others saw the need for a broader approach in which the trade-distorting potential of a national grains policy would be constrained by international agreements on support levels. When the Community offered those alternatives through its *montant de soutien* and self-sufficiency ratio proposals, the United States withdrew. This was only partly due to the fact that the levels of support and self-sufficiency that the Community offered to bind were too high. More important was the fact that the US was not prepared to apply these generic approaches to disciplining national farm programmes and organizing world agricultural trade to itself. The United States 'had no discernible master strategy . . . for negotiating domestic policies partly because . . . [it] never considered this to be a realistic possibility'.[14] Had serious attempts been made to negotiate agricultural trade arrangements that involved interfering with established US farm programmes, it is highly probable that they would have been repudiated by Congress.

In view of subsequent developments in European agriculture under the CAP, with the benefit of hindsight, it may be thought that the United States and the other exporters made a strategic error in not seizing Europe's offer to limit the support it gave its farmers through commodity programmes and to limit domestic producers' market shares. Indeed, one senior American negotiator said as much shortly after the Kennedy Round was concluded (Hedges, 1967, pp. 1332–41).

For Europe, the agricultural component of the negotiations was bound to be difficult for, at France's and Germany's insistence, the fundamental internal decision had been made to follow a protectionist agricultural policy in exchange for internal and external free trade in manufactures. During much of the early stages of the Round the Community did not have an agricultural policy to negotiate about. Once agreement had been reached in Brussels, it was too fragile to permit significant changes to be made in the Geneva negotiations lest the European construction fall apart. Quite the reverse – the French and the Commission used the negotiations in Geneva as an accompaniment to the battle they were waging in Brussels to get agreement on the commodity marketing arrangements, common prices and joint financing of the CAP.[15] The Community spent the 1960s giving form and substance to its first common policy. It could not respond positively to demands that it reduce the protection it afforded. For this reason, an even more sustained attack by the United States on Europe's common agricultural policy would likely also have been doomed to failure.

In the final analysis, despite its rhetoric and its initial attempt to link the liberalization of trade in manufactures to progress in agriculture, the United States did not have the will to press to the limits its agricultural case against the Community. Agriculture was the first common policy of the Community. To insist on the dismantlement of the CAP might have jeopardized the goal of European unification that the United States had promoted for two decades. The mitigation of the CAP's external effects was not high on the policy agenda of either the Six or the US Department of State in the 1960s.

5 The Tokyo Round[1]

1 INTRODUCTION

Although the Kennedy Round had accomplished much, even more remained to be done. There were still a large number of high tariffs in most country schedules. Many non-tariff measures still distorted trade. Little had been accomplished for agriculture. Looking ahead, the entry of the UK into the European Economic Community and the spread of its growing web of preferential trade agreements, the rapid industrialization of a set of middle-income countries, the demands of the less-developed countries for a better place in the international trading system, and the growth of economic transactions with the centrally planned economies, were all developments which pointed to the need to re-examine the design of the international trading system in relation to changes in the world political and economic order. More generally, the creeping protectionism observed after the conclusion of the Kennedy Round indicated the conflict between the deepening economic inter-dependence of nations on the one hand and their desire for policy autonomy and their reluctance to adjust to changing patterns of competition on the other. All this pointed to the need to refurbish the GATT in ways that would extend its domain and re-establish its authority.

At the conclusion of the Kennedy Round, the CONTRACTING PARTIES recognized the importance of maintaining the momentum of trade liberalization. There was a presumption that a new round would be launched at some future date, and a Programme of Work – which was regarded as technical preparation for a future round of trade negotiations – was initiated at the 24th session in November 1967 (BISD 15S/67–74). All this work was geared towards the identification of the problems which needed to be addressed, the assembly of factual information about them, and, where possible, the identification of possible solutions.

The Programme of Work was organized around three committees. The Committee on Trade in Industrial Goods focused on the problem of tariff disparities and on identifying and classifying national non-tariff measures that impeded or distorted international trade. The Committee on Trade and Development, with advice from a newly established

Special Group on Trade in Tropical Products, examined the special problems of the developing countries in the trading system. The Committee on Agriculture explored the options for achieving the GATT's objectives in trade in agricultural products. This Committee made an effort to identify and quantify the effects of government measures affecting production, imports, and exports of eight major temperate zone and tropical agricultural products. Proposed 'mutually acceptable solutions' were also discussed.

The orderly process of preparing for a major new round of trade negotiations was overtaken by the explosive release of the tensions that had been gathering in the international monetary, trade and development systems for some time, and the sudden emergence of new crises in the global energy and food systems. The trade negotiations that were conducted in the 1970s became part of a series of interrelated negotiations on reforming the international economic order in the context of a changed global environment.[2]

The search for fundamental changes in global economic systems began in the international monetary system. The United States took the initiative that compelled the international community to confront the gathering infirmities in the international economic order. In August 1971, without consultation with other countries, US President Richard Nixon and Treasury Secretary John Connolly announced that the United States would no longer convert dollars into gold at the official price to settle international transactions. Par values and convertibility of the dollar – the foundation of the Bretton Woods system – ceased to exist. To focus attention on the trade side of the US' balance of payments problem, a 10 per cent surcharge was imposed on imports, and the US demanded immediate bilateral negotiations on changes in the trade policies of the European Community, Japan, and Canada, pending wider multilateral negotiation of reforms in the international trading system.

A temporary resolution of the monetary crisis was achieved in December, 1971 in the Smithsonian Agreement, under which the dollar was devalued against major currencies and a temporary regime of new central rates and wider margins was established. The Smithsonian Agreement called for fundamental reform of the international trading system in parallel with reform of the international payments system. Intensive bilateral US–EEC and US–Japan consultations were held in the first weeks of 1972. The resultant Joint Declarations on International Trade Relations, which were conveyed to the GATT Secretariat in February 1972, amounted to an agreement between the three dominant powers in the world economy to begin a new round of comprehensive trade negotiations. The US–EEC declaration recorded that agreement had been reached

> to initiate and actively support multilateral and comprehensive nego-
> tiations in the framework of GATT beginning in 1973 . . . with a
> view to the expansion and ever greater liberalization of world trade . . .
> through the progressive dismantling of obstacles to trade and the
> improvement of the framework for the conduct of world trade. . . .
> The negotiations shall be conducted on the basis of mutual advan-
> tage and mutual commitment with overall reciprocity, and shall cover
> agricultural as well as industrial trade.

The joint US, EEC, and Japanese call for a world trade conference was welcomed by the GATT Council at its meeting in March 1972, and a large number of countries announced their support for a new round of comprehensive trade negotiations. The formal decision to initiate preparations for a new round – to begin the following year and to be completed by 1975 – was taken by the CONTRACTING PARTIES in November 1972 (BISD 19S/12–3). A Preparatory Committee was established to clarify the issues that needed to be addressed, and to try to identify the working hypotheses that could form the basis of a comprehensive agreement. The ceremonial launch of the multilateral trade negotiations (MTN), the seventh GATT Round, occurred in September 1973 at a special meeting of ministers convened at Tokyo.[3]

The preparatory and early stages of the comprehensive trade negotiations were overlapped by negotiations on other issues in other fora. The MTN became the forum in which the trade-related aspects of these other 'world order issues' were addressed. Components of the MTN were intimately related to aspects of the world food and energy crises that developed in the first half of the decade, and to the demands of the less developed countries for 'a new international economic order' that were addressed in the mid-1970s.

The world food crisis began in the period of the run-up to the Tokyo Round. Its immediate causes were the coincidence of three factors – the decision of the US and Canada in 1970 to run down grain stocks and reduce cropped acreage, a shortfall of 3 per cent in world cereals output between the 1971 and 1972 harvests, and the decision of the USSR in 1971 to import grain to offset domestic production shortfalls. Together, these forces triggered a surge in world grain prices. Supplies were further depleted and prices raised still higher by a 4 per cent reduction of world grain production between the 1973 and 1974 harvests, a failure of the Peruvian anchovy harvest, and continued large volumes of grain imports by Eastern bloc countries. The results included food-led price inflation and dislocations in animal agriculture in Western countries, an increase in the cost of commercial grain imports in food importing countries, a reduction in food aid shipments to countries dependent on concessional supplies, and the intensification of hunger and malnutrition among poor people in many parts of the world.

The functioning of the global food system became a world order issue. The principal forum in which its perceived inadequacies were addressed was the World Food Conference held in Rome in 1974.[4] However, aspects of the international food crisis became part of the agenda of the multilateral trade negotiations which had begun in Geneva. These included the role of international commodity market management in increasing the stability of world food markets, the creation of trade arrangements that would ensure equitable sharing of the burdens of adjusting to changing market conditions and providing food aid to developing countries, the imperatives of ensuring that the less developed countries could earn enough foreign exchange from their exports to meet *inter alia* their increased food import bills, and the need to place under stronger international disciplines the export restrictions that some countries imposed to offset food price increases in their national economies.

In October 1973, a month after the MTN were launched, in the aftermath of the Arab–Israeli war, the members of the Organization of Petroleum Exporting Countries (OPEC) quadrupled the price of oil. The energy crisis fuelled inflation, weakened growth, disrupted payments balances and caused huge international redistributions of capital. Oil-importing developing countries were particularly hard hit, and the borrowing that they were forced to undertake in the 1970s to finance their imports of fuels and other essential consumption and development goods became the genesis of the debt crisis of the 1980s. The worsening balance of payments and debt positions of the developing countries made improving their position in the international trading system a priority goal of the MTN.

Much of the discussion of the place of the less developed countries in the world economy was conducted in the United Nations in parallel with the MTN. The UN had declared the 1970s to be the second Development Decade, and a resolution on the creation of 'a new international economic order' (Djonovich, 1978, pp. 527–9) was passed at the close of the sixth special session of the UN General Assembly held in April and May of 1974. Thereafter, much of the formal debate and negotiations on giving effect to this concept – including the attempt to devise an 'integrated programme for commodities' – was conducted, during the 1974–6 period, under the auspices of the fourth United Nations Conference on Trade and Development (UNCTAD IV) where the developing countries had the support of a committed secretariat and a voting majority.

2   OVERVIEW OF THE TOKYO ROUND

Problems and Players

At the end of the Kennedy Round the international trading system was showing important systemic weaknesses. The GATT's rules and agreements were everywhere being tested by the protectionist pressures that were driven by weak economic performance, misaligned exchange rates, payments imbalances, and the difficulties of making the structural adjustments necessitated by the emergence of new international competitors.[5]

Some of the central tenets of the GATT system were being violated or ignored. The proliferation and enlargement of customs unions, free-trade areas, and preferential regional association agreements were unconstrained by the principle of nondiscrimination or the specific requirements of Article XXIV of the Agreement. Tariffs had been much reduced, but protection was being widely provided through government interventions with a sectoral focus using such non-tariff measures as regulation and subsidies, about which the General Agreement was weak or said little. In an era requiring structural adjustment to changing patterns of comparative advantage, countries were circumventing the temporary protection or safeguard provisions of the General Agreement (notably Article XIX) by using such extra-legal measures as 'voluntary export restraints' (VER) or 'orderly marketing agreements' (OMA) to limit the flow of competitive imports into their markets. In the rich countries the progressive erosion of the normative framework was reflected in new language which spoke of 'trade reciprocity', 'trade remedies', 'managed trade' and a 'results-oriented trade policy' in which negotiating progress should be measured against 'numerical targets'. Few developing countries applied or observed the Agreement, and collectively they were demanding extension and codification of their special privileges and differential obligations under its provisions. The dispute settlement procedures of the General Agreement were not working well even as compliance with its provisions faltered. Whole areas of international trade – notably agriculture, textiles, and apparel – were not covered by adequate GATT rules and disciplines, and the conditions of commerce in these areas were the antithesis of an orderly and liberal trading regime. As national economic policies became more protectionist, trade relations between countries were becoming more fractious.

As in monetary reform, leadership in advocating wide-ranging reform of the international trading system fell to the United States. President Johnson had asked his Special Representative for Trade Negotiations, William Roth, to prepare a report on the issues that should be addressed in future trade negotiations (Roth, 1969) and he encouraged Congress to examine America's trade policy in the broader context of its future international economic relations (US Congress, 1970). Strengthening the GATT and liberalizing trade were consistent with the commitment of the US to open markets, clear rules and juridical remedies; and opening foreign markets, or 'export politics', was the traditional means by which the US Executive and Congress resisted domestic protectionist pressures (Destler, 1986). Furthermore, whereas in an earlier era the US had been willing to tolerate discrimination and protectionism by Europe and Japan as a step to forging a strategic alliance among the liberal and affluent democracies, by the 1970s the US was more interested in having these superpowers assume a share of the costs, commensurate with their stake and their abilities, of maintaining the international systems of trade and payments, absorbing imports from the newly industrialized countries, and providing foreign aid.

For the United States, as the decade began, agricultural trade reform was of crucial importance. Agriculture was one of the few major industries that still favoured freer trade, and Congressional support for the legislation that would authorize the Administration to embark on a new round of trade negotiations would only be forthcoming if it promised to deliver expanded export opportunities for US farm products. Agriculture had been identified in two reports as being an industry in which the US had a clear comparative advantage, but where this advantage was being frustrated by the agricultural and trade policies of other countries (Petersen, 1971, and the Williams Commission, 1971).

At the beginning of the 1970s, neither Europe nor Japan was enthusiastic about global trade negotiations. The political energy of the European Community was directed towards enlargement by the accession of the United Kingdom, Denmark, Ireland, and Norway,[6] and creating association arrangements with other countries in Europe and the Mediterranean basin and with the former colonies of the metropolitan powers. European leaders knew that the Common Agricultural Policy would come under attack in any comprehensive trade negotiations. The creation of the CAP had been extremely difficult, and it was still not completed. Yet it was imperative to withstand assaults on this first common economic policy lest the whole economic and political construct that was 'Europe' collapse. Beyond this, the Community was still an embryonic institution. It was too loose an association to have the 'world order vision' or the 'milieu goals' that Americans were prone to proclaim and demand, incapable of adopting a coherent approach to the interrelated aspects of international economic relations that the Smithsonian Agreement had promised, and poorly equipped to adopt a flexible approach to the multifaceted agenda of a comprehensive trade negotiation.[7] The Community had a propensity to externalize its internal conflicts, a preference for bilateral and sectoral arrangements, an affection for discriminatory preferences, and a disinclination to support effective international dispute settlement mechanisms.[8] At most, going into the MTN, the Community might have welcomed a modest tariff package on industrial goods, and even this might have been resisted by the French – who had quite enough competition from German industry – and by those in Europe who saw the common external tariff of the customs union as a unifying force.

As a major trading power whose seemingly uncontainably competitive exports were subject to trade barriers, discrimination, and safeguard actions in other developed country markets, Japan had much to gain from a successful trade negotiation. But, because of its mercantilist ethic, consensual politics, and system of governance in which the bureaucracy defended constituency interest, Japan found difficulty in playing an activist role in the process of reforming the world trading system. As Ostry observed, its preference for 'quietist inertia' and its reactive style were expressed in the defence of national trade interests rather than the development of global trade policy (Ostry, 1990, p. 37). In agriculture, having become a large and regular, if regulated, importer of many other farm and food products, Japan felt that few changes were needed in its agricultural import policies. And having the economic means to take what it wanted from world food markets, Japan did not have a compelling interest in world food security issues.[9]

For all these reasons, the negotiating focus of the MTN was on the United States and the EEC, with the former acting as the demandeur. The other economic superpower, Japan, was never a third force. The remaining developed countries played a subsidiary role and did not act as a bloc. Capitalizing on the uncertainties that accompanied the turmoil in commodity markets and concerns about the limits to global growth that were a feature of the early 1970s, the developing countries succeeded in making their demand for an improved position in the trading system a subject of the negotiations. However, once it became clear that their unity and coercive power were illusory, their influence in the MTN and the willingness of developed countries to meet the developing countries' trade demands both waned.

The Agenda

The broad goals of the MTN – to liberalize trade, to extend the authority of the GATT to non-tariff measures, to strengthen its rules, to improve its dispute settlement procedures, and to harness trade to development – translated into a prolonged negotiation on a large number of issues which were pursued in a negotiating structure of great complexity. However, five specific objectives were pre-eminent: reducing tariffs on manufactures; devising new rules to reduce the trade-distorting effects of non-tariff measures; strengthening the system of temporary safeguards; improving the position of the developing countries in the international trading system; and reducing the trade effects of agricultural support and protection.

The binding and reduction of tariffs – the traditional business of the GATT – was a goal both in its own right and also as a means of reducing the discriminatory preferences inherent in the formation of regional trading blocs. However, a method of reducing tariffs had to be devised that would also deal with the disparities in the profiles of country tariff schedules. It was eventually agreed to use a formula approach under which the depth of tariff cut was proportionate to the height of the existing tariff.[10] If applied without exception this would have resulted, at the end of the 8-year phase-in period, in an average cut of 40 per cent in the tariffs levied on manufactures by the developed countries and a measure of tariff harmonization between them.

The work programme initiated in the GATT at the end of the Kennedy Round had assembled a list of some 800 examples of 27 distinct types of national non-tariff measures (NTM) that distorted trade. An attempt had to be made to discipline the existing measures with the most egregious trade effects, and to create a framework of obligations which would encourage governments to choose trade-neutral instruments as they deepened their involvement in economic life. The task was lent urgency by the fear that, as tariffs were lowered, governments would increase protection for national industries through subsidies and regulations that favoured domestic firms.

The task facing the negotiators was to agree on substantive rules governing the use of trade-affecting subsidies and technical standards, which would ensure fair competition between domestic and foreign firms, and to establish appropriate surveillance, complaints and dispute settlement mechanisms. Because of difficulty in getting unanimous agreement to amend the articles of the General Agreement, the approach used was to write codes of good conduct.

The right to impose temporary restrictions on fairly traded but intolerably disruptive imports of particular products is a normal feature of trade agreements. Temporary import restrictions or 'safeguard' actions were sanctioned by the General Agreement, notably by Article XIX. A well-designed safeguards system is economically beneficial in that it encourages market openness. In some respects Article XIX was too lenient in that it did not define serious injury, placed no time limits on safeguard action, did not require adjustment assistance for the protected industry in anticipation of the removal of protection, and was not subject to prior international approval and adequate surveillance. In other ways the article was too strict: the requirements that the trade restriction be applied without discrimination against all suppliers, and that the importing country provide compensation or risk retaliation, encouraged countries to deal with 'disruptive' imports by resorting to VER or OMA instead. By this means the restricting country avoided the need to prove injury or provide compensation, and trade restrictions could be targeted at the 'offending' suppliers. Reform lay in the enactment of a new code on the use of temporary import restrictions or safeguard measures that would prevent abuse by providing more precise definitions, more explicit rules, and effective international surveillance.

The Tokyo Declaration had agreed that the MTN would attempt to provide additional benefits for the developing countries so as to achieve a substantial increase in their foreign exchange earnings. This was to be accomplished *inter alia* by securing an improvement in the conditions of access to rich-country markets for the products of export interest to the developing countries. Trade in tropical products was to be treated as a special and priority sector. This meant lowering the tariffs, taxes, and non-tariff barriers faced by developing exporters, and paying particular regard to reducing the high levels of effective protection to value-adding activities provided through tariff escalation. However, the developing countries also sought additional trade benefits that would accelerate their development by agreements that extended and enshrined the principle of their 'special and differential' treatment. Measures that would have this effect included a standing legal basis for the generalized system of preferences that had been introduced in the 1960s, the binding of the margins of preference these and other arrangements provided, laxer standards for developing countries under codes of trade conduct, and further relaxation of the exceptions they enjoyed under the articles of the GATT in recognition of their balance of payment situations and development needs.

In the period leading up to the opening of the Tokyo Round, agriculture held its customary place at the centre of the trade conflicts between the United States and the European Community, and was a serious trade problem among the developed countries in general. All the influential trade policy analyses of the day had no hesitation in singling out agricultural protectionism as the most difficult trade policy problem dividing the developed countries, and its resolution as a priority task for, and a measure of the success of, the negotiations which were about to be launched.[11] US spokespersons affirmed that fundamental reform of the conditions of agricultural trade was an imperative.[12] Europe agreed to the inclusion of agricultural trade on the MTN agenda only grudgingly, and in guarded language which indicated that trade in the products of this sector could not proceed along the same path and at the same speed as trade in manufactures. However, by the time substantive negotiations opened in early 1975 the world was in the midst of a food crisis.

Accordingly, while the objective of improving access to import markets did not disappear from the agricultural trade agenda, in the all-important grain sector – always the focal point of the agricultural trade conflicts between the US and the EEC – it was joined by the additional objectives of joint action to reduce price volatility in world grain markets and adapting trade arrangements to counter world hunger. The desire for stability resonated with the EEC's stance that stabilization of international agricultural markets must be a precondition for, and an accompaniment of, their liberalization and that this required the 'organization' of markets. The possible need for intergovernmental cooperation to prevent disorder in world food markets caused the agricultural trade negotiations in the MTN, particularly as they related to grains, to intersect with the negotiations on enhancing world food security and fashioning an integrated programme for commodities that were being conducted in other fora.[13] In the lexicon of the day, international cooperation for 'enhancing market stability', the exercise of 'concerted disciplines' to enhance 'world food security', and 'equitable burden sharing' in these tasks joined the traditional concern with improving 'access to markets' in the substantive agricultural trade objectives of the MTN.

The grain exporters were, of course, anxious to see support and protection lowered on a permanent basis in anticipation of a turnaround in market conditions, but this future possibility carried little negotiating weight. Similarly, although the price collapse and increased trade protection in world markets for meats, which were the 'flip side' of the surge in grain prices, made the conditions of access and competition even more urgent for the commercial exporters of these products, individually and collectively they lacked the political and economic weight required to keep their concerns about import protection and subsidy competition at the forefront of the negotiations.

Timetable and Structure

The Declaration that launched the Tokyo Round (BISD 20S/19–22) set out the aims and objectives of the negotiations, the principles on which they were to be conducted, and the topics to be included. The immediate objectives of the agricultural exporters were met by agreement that the negotiations 'shall cover . . . both industrial and agricultural products'; the EEC's reservations were represented in the observation that 'as regards agriculture, an approach to negotiations [shall be adopted] which . . . should take account of the special characteristics and problems of this sector'.

A Trade Negotiations Committee (TNC) was created to elaborate and put into effect detailed negotiating plans and to supervise the progress of the negotiations. In February 1974 the TNC established six subcommittees (later called groups) corresponding to the subjects on which it had been agreed to negotiate – tariffs, non-tariff measures (NTM), the safeguards system, sector measures, tropical products, and agriculture. With the reluctant agreement of the United States, in May 1975 Group Agriculture established subgroups for the major internationally traded temperate commodities, grains, manufactured dairy products, and meat, to deal with 'all elements relevant to trade in these three products'.

The groups and subgroups initially concerned themselves with updating and completing the data collection and analysis begun under the 1967 work programme. Most of 1974 and 1975 was spent in exchanging data and developing negotiating positions. Later, the groups and subgroups became the location of concrete negotiations – the specification of the problems to be solved, the exchange of views on how these might best be addressed, agreement on 'working hypotheses', the tabling of concrete proposals, the preparation of draft texts and requests and offers, and the initialling of agreements.[14] Some of this work was recorded in notes sent to and distributed by the GATT secretariat, but many of the important 'non-papers' that record this work never reached the public domain.

Once the negotiating structure was in place, it ground to a halt. The MTN were to 'wallow in aimless debate' for the next three and a half years, until July 1977. Progress was held up by two main factors. After the experience of the Kennedy Round, no country was willing to negotiate with the United States without assurance that agreements reached in Geneva would not be thwarted in Washington by Congress.[15] Consequently, substantive negotiations could not begin until the US Administration had secured a suitable negotiating mandate from Congress. President Nixon sent a Trade Reform Act to Congress in April 1973 but it was not signed into law (by President Ford) as the Trade Act 1974 until January 1975. It included a 'fast-track procedure' under which Congress empowered the Executive to enter into agreements on non-tariff measures and made acceptance of the necessary implementing legislature subject to a time-limited 'up or down' vote – that is, Congress had to accept or reject the negotiated package as a whole without power to delay or amend.[16]

Secondly, progress in the negotiations as a whole was also prevented by a dispute over the competence of Group Agriculture. The US Trade Act required that 'to the maximum extent feasible, the harmonization, reduction and elimination of agricultural trade barriers and distortions shall be undertaken in conjunction with the harmonization, reduction and elimination of industrial trade barriers and distortions'. This was the initial basis of the US' approach to negotiating on agriculture – that there should be no separate negotiations about factors limiting and distorting trade in the products of this sector.[17] Once Group Agriculture was formed, the US argued that trade in agriculture was discussable in all groups, specifically in the groups and subgroups dealing with tariffs and non-tariff measures, and that whatever was agreed there was, in principle, applicable to trade in farm and food products. The EEC insisted that trade in agricultural products be dealt with only within Group Agriculture. The Community was not prepared to have its policy freedom in agriculture, and the highly specific commodity provisions of the CAP, constrained in uncertain ways by multilateral formulae and codes of general applicability, and was determined to have agriculture dealt with in a single forum where it could better control the expected assault on the CAP. Furthermore, as the US Administration did not have a clear political mandate to offer concessions for the agricultural products it imported,[18] the United States sought to exchange agricultural trade concessions for offers in manufactures and changes in its trade remedy procedures. Hence, from the EEC's perspective the negotiating leverage of the US in agriculture was minimized by sectoral negotiations.

The issue was fudged in May 1975 when the US conceded that the agricultural subgroups would 'concern themselves with the agricultural aspects of tariffs and non-tariff measures' and the EEC agreed that these would be considered 'in conjunction with the work of the Tariff and Non-tariff Measures Groups'. The US made a further concession in December 1975 when it agreed to the EEC's demand that products not covered by the three commodity subgroups would also be handled in Group Agriculture.[19] But this was still not sufficient to bridge the fundamental differences on agriculture between the two chief protagonists, and the negotiations continued to languish. Though work continued in a desultory fashion in the commodity subgroups, Group Agriculture ceased to meet. The impasse on agriculture held up progress in the other groups. The stalemate lasted until July 1977.

In truth, while the differences on agriculture were serious, the MTN as a whole did not hold centre stage in the 1973–7 period. Political attention was pre-empted by a whole range of events and issues of greater urgency and larger geo-political significance which required the attention of world statesmen – the Yom Kippur War and the oil crisis it triggered, world hunger, North–South relations, and the beginnings of détente. And trade policy had to share the stage with continuing monetary instability and balance of payments crises, and with deepening inflationary and recessionary trends. On top of all this, US leadership was impaired by the Watergate crisis, President Ford's interregnum, and preparations for the 1976 US presidential election. There were also important elections in Germany and Japan in 1976. Consequently, it was not until after the Carter Administration took office in January 1977 that the political conditions existed for negotiations proper to be pushed forward.

One of President Carter's first foreign economic policy initiatives was to secure a commitment to complete the negotiations from the heads of governments assembled for the economic summit in London in May 1977. The men he appointed as Special Trade Representative and Secretary of Agriculture, Robert Strauss and Robert Bergland, had lower expectations about what could be accomplished for agricultural trade than their predecessors, Frederick Dent and Earl Butz, nor were they as ideologically opposed to the 'commodity arrangements' approach that the Community was advocating in Geneva. In mid-July 1977, at a meeting between Strauss and EEC Commissioners Wilhelm Haferkamp, Finn Olav Gundelach and Etienne Davignon, the US capitulated to the EEC by stating that the MTN would not 'upset any structural agricultural policies of the EEC', and that because agriculture was 'different' it could be negotiated 'in parallel' with industry. In return the US obtained a definite timetable for completing the Tokyo Round, and an understanding that there would be 'a substantial result' for agriculture.[20] With the political push given by summit leaders, and the break in the agricultural log jam by the agreement that trade talks in agriculture and industry could proceed simultaneously but under different rules, substantive negotiations were able to proceed in all the groups.

Agreement was soon reached on the key 'working hypotheses'. Tariffs on manufactures would be reduced and harmonized through the use of the Swiss formula. Other trade restrictions on industrial products would be handled through requests and offers. Non-tariff measures would be subject to GATT disciplines through codes, some of which were at an advanced draft stage. Later that year it was formally agreed that the Swiss formula for tariff reduction would not apply to tariffs on agricultural and food products. Access for agricultural products would be improved by a request and offers procedure. Requests were to be tabled by 1 November 1977, and offers by 15 January 1978. At this stage, no agreement had been reached on the role and nature of commodity arrangements, nor on the degree to which non-tariff measure codes would apply to agriculture.

The pace of negotiations quickened further following a meeting in January 1978 between US Special Trade Representative Strauss, EEC Commissioner for External Affairs Haferkamp and Japan's Minister for External Economic Affairs Ushiba. At the economic summit held in Bonn in mid-July 1978, it was possible to report substantial progress (GATT, 1978). The heads of governments made a further commitment to conclude the detailed negotiations in all areas by 15 December 1978. This was now urgent since the 'fast-track' authority of the US Trade Act 1974 expired on 3 January 1980, which in practice meant that negotiations had to be substantially completed by March 1979.

Completion of the negotiations was delayed by difficulties in reaching agreement on balanced packages of bilateral access offers, on the contentious clauses of the codes, and on the special provisions for the trade of the developing countries. However, by April 1979 it was all done. The Tokyo Round came to a quiet conclusion at an anticlimatic meeting of the Trade Negotiations Committee, 11–12 April 1979, with the opening for signature of a *procès-verbal* and the texts of the agreements on tariffs concessions, six codes of conduct in the use of trade-affecting non-tariff measures,[21] a framework agreement for trade of the developing countries, and arrangements for dairy products and beef.[22]

3 TASKS AND PROPOSALS IN AGRICULTURE

In the run-up to the launch of the Tokyo Round there was no doubt about the nature of the agricultural trade problem. As the Assistant Director-General of the GATT put it in 1970:

> The situation of world trade in temperate agricultural products is progressively deteriorating. An impressive array of national and regional agricultural policies, buttressed on the one side by restrictions on imports and on the other by incentives to exports, are slowly closing commercial markets to exporters. The Kennedy Round failed to end this process or set bounds to its eventual spread. (Patterson, 1971, p. 631)

All could agree that the GATT had failed to come to grips with the underlying agricultural policies, and had little success in dealing with their secondary effects through Articles XI and XVI. Nor was there any doubt about the direction in which a fundamental solution lay:

> If the restrictions on international farm trade are to be diminished rather than expanded during the 1970s, domestic farm policies must be accommodated to the terms and conditions of a liberal trade policy. It is as simple as that and there is no other magic answer. (Johnson, 1970, p. 400)

To this end, the US had proposed in the GATT Committee on Agriculture 'that governments agree to accept as the ultimate objective, that all quantitative restrictions, variable levies and related restrictive measures be removed and replaced by fixed tariffs at non-prohibitive levels, and that all government aids to exports, direct and indirect, should be eliminated' (Patterson, 1971, p. 636). At the rhetorical level, this continued to be the US' negotiating objective, but emphasis on it was diluted by the change in market conditions that occurred during the period of the negotiations.

Driven by supply shortfalls and new demands from the centrally planned economies and developing countries, the value of world trade in grains and oil-seeds began to soar even as the MTN got underway. For the US, the ability to continue a high level of grain and oil-seed exports was important to the balance of payments and to the value of the dollar, and to maintaining farm incomes and asset values while 'getting government out of agriculture' and continuing to 'plant hedgerow to hedgerow', and avoiding the need to revert to the costly farm support programmes of the past.[23] International action to stabilize markets, including grain markets, was not a high priority in commercial terms.[24] On the other hand, by 1974, enhancing global food security – which had market stabilization, grain reserves, production adjustment and food aid dimensions – had acquired a foreign policy significance,[25] and the US was very interested in having the other rich countries bear an equitable share of the costs of the international programmes which were then being advocated. Additionally, the US and other exporters were anxious to have Europe share the burdens of adjusting production and consumption in response to changing market conditions.

To many, all this pointed to the need for international cooperation to improve the functioning of world commodity markets, to be handled as one of several goals within a wider framework of measures for commodity market 'management'. This was exactly what the EEC said it was offering. The EEC's position on the treatment of agriculture had been set out in the 'overall approach' to the MTN (Council of the European Communities, 1973). The Community's negotiating offer was based on three pillars. First, Community authorities maintained, and appeared to believe, that agriculture as a sector was fundamentally different because of its social, economic and political importance, the universality of government intervention,[26] and the chronic propensity for farm product markets to be unstable. Second, nothing in the MTN should be allowed to disrupt, still less dismantle, the CAP. Third, multilateral cooperation to stabilize markets was a precondition for their liberalization. These propositions were combined by the Council in the overall approach in the following words:

> The Common Agricultural Policy corresponds to the special condition of agriculture within the Community. Its principles and mechanisms should not be called into question and are therefore in no way a matter for negotiation. ... The specific objective of the agricultural negotiations should be to secure expansion of trade in stable world markets, in accordance with existing agricultural policies. ...

The conditions for the expansion of trade would be more favourable if the stability of world markets were better assured. The best way of achieving that objective would be to conclude appropriate world-wide arrangements.

The Community proposed that two types of intergovernmental commodity agreements be negotiated. First, it envisaged price disciplines for some products. For storable commodities – each of the major grains and sugar – agreements were suggested in which prices were held within a negotiated range by varying the size of the stocks held by importers and exporters as prices passed through 'trigger' levels. There would also be preferential purchase and supply obligations at the agreed price minima and maxima. For products for which stockpiling is not a normal instrument of commercial management – specifically, the more homogeneous dairy products, butter, and milk powders – it proposed that exporters agree not to sell at prices below agreed minima. Exporters would also give supply preference to importing signatory countries at agreed price maxima.

Second, for other agricultural products – of which the most important were bovine meats[27] – the Community proposed 'joint disciplines' in which exporters would ensure that their exports flowed to market in an 'orderly manner ... as regards both trade and price policies' and enter into agreements with importers 'to sell in conditions compatible with a satisfactory development of the internal market of the importing country'. What this meant was never entirely clear, though at best it might have entailed importers' assuming minimum access obligations in return for some agreement on the use of export subsidies.[28] There were also proposals for creating a mechanism for sharing information on market conditions and prior consultations on policy changes and trade actions, including safeguard measures.

The US and the smaller commercial agricultural exporters kept asking the Community what its proposals on intergovernmental commodity agreements had to do with agricultural trade liberalization. The reply was always that the need to 'respect' such matters as minimum export prices and stockholding commitments would force all countries to adapt their national farm programmes and trade arrangements to conform with their international obligations. Further, there were suggestions that stable markets could be 'shared'. Such indirect influences on the level of agricultural protection and subsidization were seen to be worth little in negotiating terms. Moreover, the exporters wanted agricultural trade to be liberalized rather than divided up by negotiation. Market sharing would forever deny them the fruits of their comparative advantage. And they were certainly not prepared to legitimize market shares that, as they saw it, had been improperly (and possibly illegally) acquired with the aid of subsidies.

Significantly, though several nongovernmental groups urged that reductions in agricultural protection and support be achieved through the use of the *montant de soutien* technique suggested by the EEC in the Kennedy Round,[29] and though studies done for the FAO had shown that the technique was practicable,[30] neither the Community nor the US ever proposed that this approach be employed. The Community's disavowal was readily explained – it had no interest in making commitments to reduce the protection and subsidization provided to Europe's farmers by the CAP. For the Nixon, Ford, and Carter Administrations, the United States' restraint appeared to reflect the strategic decision to try to reform agricultural trade by having it treated in the same manner as industry in respect of both access and codes.[31]

4 THE RESULTS FOR AGRICULTURE

The agreements affecting trade in temperate zone agricultural products reached in the Tokyo Round cover three areas: access, commodity arrangements and the codes on subsidies and standards.

Market Access and Commodity Arrangements

In the final weeks of the MTN, a great deal of effort went into securing bilateral agreements which, when stitched together, would add up to a balanced package of access concessions and benefits.[32] Improved and assured access for agriculture was achieved by a request and offers procedure, conducted between pairs of countries, on a product-specific basis. The bilateral access negotiations were conducted within Group Agriculture. Access was improved by both tariff reductions and bindings and by the establishment, enlargement or binding of import quotas. The Swiss formula used to lower and harmonize tariffs on manufactures was not employed. Though negotiated bilaterally, the reduced tariffs were, of course, applied on a nondiscriminatory basis. Many quota increases, on the other hand, favoured specific suppliers.

Tariff concessions were made on a quarter of the dutiable agricultural and food items entering nine developed countries – on $12 billion of imports out of a total of $48 billion. On items on which concessions were made, the average tariff cut amounted to 40 per cent. Tariffs were reduced or bound on hundreds of items in the tariff schedules of the negotiating countries. The extent to which agricultural trade was actually liberalized as a result of these tariff concessions is unclear. Many items on which tariffs were reduced remained protected by other border measures and supported by domestic programmes, and some of the bound tariff rates were higher than the duties actually being charged. The benefits of quota changes were generally more tangible and were likely to be more important in economic and trade terms. In particular, the restrictive import quota regimes for beef and dairy products operated by the EEC, the United States, Japan, and Canada were marginally eased and limited access commitments were granted, mainly, but not exclusively, to the benefit of Australia and New Zealand.

Discussion of the EEC's proposal for the negotiation of intergovernmental arrangements for the three major traded commodities was initially taken up in the three subgroups of Group Agriculture.

Grains

The Ford Administration had maintained that while it did not rule out international commodity agreements, it was sceptical of their value. It did not participate in substantive negotiations on the matter in its term of office. The Carter Administration which took office in 1977 was more willing to explore this vehicle. However, the US had no interest in an arrangement for feed grains, and argued that arrangements for wheat, the principal food grain, should be directed toward the negotiation of a new International Wheat Agreement (IWA) in the International Wheat Council (IWC).[33] This was seen as the appropriate forum in which to negotiate an arrangement with stabilization, food security, and burden-sharing objectives, leaving the issues of access and subsidies to be addressed in the GATT.[34]

Negotiations on an IWA with economic provisions were conducted in the IWC, under the auspices of the UNCTAD, in a 60-week period between late 1977 and the early weeks of 1979.[35] The central proposal envisioned international wheat price management through the coordinated acquisition and release of nationally owned stocks. Disagreements arose over every substantive provision – the overall price range to be defended, the role of price and quantity 'triggers' for stock and policy actions, the price or quantity points at which agreed action should occur, the size and shares of reserves, food aid commitments, and the treatment of the developing countries. The negotiations were abandoned in February 1979, though the 1971 IWA, which had no economic provisions, was extended until 30 June 1981.

The GATT MTN grains subgroup ceased to meet in Geneva while the negotiations on wheat in London were in progress. When these were terminated, there was neither the time nor will to accomplish much on access. Very few specific access concessions were negotiated for the major grains. Additionally, discussions which had been initiated at the request of the EEC on a consultative arrangement for coarse grains were dropped when the negotiations on wheat failed.

The urgency of reaching agreement may have been reduced by the fact that by 1977–8, following a series of good harvests, the US was again unilaterally idling cropland and building stocks. However, even if prices had been stable over the period, the diverse and inconsistent objectives of the participants would have made agreement difficult. The EEC wanted a stable framework in which to conduct the CAP and a high IWA price floor would have lowered the profile of European grain support prices. Commercial exporters wanted an agreement that would force the rest of the world to adjust national price policies and stocks to changing international market conditions and share more of the costs of providing food aid to developing countries. In ways that were never clear, the wheat exporters wanted an IWA with provisions on prices, stocks, and food aid to be 'folded into' an agreement in the GATT on market access and subsidies. Rich importers wanted supply assurance. Poor importers wanted this, protection against high prices, and an expanded volume of food aid maintained without regard to its opportunity cost. As the US chief negotiator later put it: 'Achieving any one of these goals in an international negotiation would have been a major task; to achieve all of them proved impossible.'[36]

Dairy Products

The fundamental problems of the world dairy industry – high degrees of protection of milk producers by most developed countries and widespread use of export subsidies, particularly by the European Community – required improvements in access to restricted markets for the genuinely efficient producers, New Zealand and Australia, and a lower level of subsidization of the production and export of dairy products. The US, Japan, and Canada were net importers of milk products with markets insulated by quantitative import restrictions. They saw little reason to make changes in their trade arrangements for dairy products that would ease the cost of dairy support to the EEC, and preferred to respond to the access problems of Australia and New Zealand through bilateral agreements. Some marginal improvements in access were granted, notably by the US and Canada for cheese,[37] and there was an expectation that the new subsidies code would place an effective curb on the use of export subsidies. However, the major vehicle supposed to improve trade conditions in the dairy sector was the conclusion of a limited arrangement for dairy products (BISD 26S/91–115).

The essentials of the International Dairy Arrangement (IDA) were proposed by New Zealand and the European Community. The agreement had two main features: minimum export prices for butter, whole and skim milk powders, and certain hard cheeses, and an institutional arrangement for information exchange and policy consultation, the International Dairy Products Council (IDPC). The minimum export prices were designed to put some constraint on the use of export subsidies and provide stronger prices for commercial exporters. The consultative mechanism was intended to encourage conflict avoidance and resolution through the exchange of forward-looking information on production, inventories, prices, and proposed changes in policies and trade arrangements. The IDPC also provided commercial dairy product exporters with a commodity-specific forum – situated in the GATT – where they could air their grievances and seek solutions to their trade problems.

In the first years after the conclusion of the IDA, world dairy product markets were firm and the provision on minimum export prices in the agreement was not tested. However, the agreement contained no enforcement mechanisms and no explicit stocking and production adjustment provisions, and while the Council could make recommendations on how problems might be resolved, this required unanimous consent and the recommendations were non-binding. Consequently, when the demand and supply conditions changed in the first half of the 1980s, there was nothing in the agreement to prevent the collapse of world prices and a reversion to subsidized export competition at prices lower than the agreed minima.

The IDA continued in force and the IDPC continued to issue annual reports on the world dairy situation and prospects, but it never had any lasting influence on the parameters of national dairy policies and the functioning of world markets for dairy products.

Beef, Veal and Cattle

From the beginning, the international agreement concerned with trade in beef concluded in the MTN (BISD 26S/84–90) was even less consequential than the IDA. Though the arrangement grew out of the EEC's vague proposals for 'concerted disciplines', it had no economic or regulatory provisions. The functions of the International Meat Council (IMC) it established were confined to reviewing the world supply and demand situation for meat and providing a forum for regular consultation on all matters affecting international trade in bovine meats. As in the IDA, the recommendations of the IMC on remedial actions to deal with evolving market imbalances or trade disputes had to be consensual, and they were non-binding. The member countries found its information exchange useful, and commercial exporters have used the IMC as a forum for discussions of their concerns. But the conclusion of the arrangement was not an important accomplishment in terms of reforming agricultural policies and trade practices in the world beef sector.

Codes

Subsidies

The Tokyo Round code of conduct on subsidies and countervail duties (BISD 26S/56–83) was the most ambitious attempt to reform the GATT since the 1954–5 Review Session. The attempt was only partially successful. Few countries were willing to have a detailed schedule of domestic subsidies with trade effects identified and exposed to explicit disciplines, and the EEC was unyieldingly opposed to the United States' main goal of extending the prohibition on export subsidies to agricultural products. This fundamental reality was symbolized by the US abandoning its 'traffic light' proposal for the categorization and treatment of subsidies.[38]

The fact that domestic subsidies can have a significant negative impact on the economic interests of other countries was acknowledged. Articles 8:3 and 11:2 of the code recognized that domestic subsidies might cause or threaten damage in three areas: 'injury' to domestic industries in importing countries; 'nullification or impairment' of expected GATT benefits by import substitution; and 'serious prejudice' to another country's interests by displacing its exports in third markets.[39] However, a definitive list of tainted domestic subsidies was not included in the code.

The code spelled out tighter obligations on the use of countervail duties, in the areas of prior consultation rights, procedures for countervail investigations, and the determination of injury and causation. Most importantly, by signing the subsidies code, the United States accepted the obligation to demonstrate injury before placing countervail duties on subsidized imports from other countries that were code signatories. Potentially, this was an important advance for other countries in handling their bilateral trade disputes with the United States. In practice, because the US implementing legislation contained only a weak injury standard and had procedural provisions that both reduced inhibitions against bringing countervail actions and made favourable rulings easier to obtain, 'process protection' in the US was unleashed. As a result, bilateral trade relations between the US and its partners on subsidy–countervail issues worsened after the Tokyo Round.

The principal new features of the GATT's disciplines on domestic subsidies were the obligation to 'seek to avoid' causing adverse trade effects with national subsidies, the right to consultations and to retaliate, and the recognition of third-market effects. These did not provide effective constraints on national subsidies, especially as they related to agriculture. No country felt the need to change elements of its domestic farm programmes to conform to obligations in the subsidy code. Thus the attempt made in the MTN to curb the adverse trade effects of national support programmes in agriculture was not successful.

In respect of export subsidies, notable accomplishments of the Tokyo Round Subsidies Code for nonagricultural trade were the expansion of the illustrative list of export subsidies, the elimination of the requirement that they result in dual-pricing, and the extension of the ban on their use to exports of minerals.

Progress for agriculture was more modest. Agricultural export subsidies continued to be tolerated under the General Agreement. The code contained no specific definition of what constituted agricultural export subsidies,[40] and no obligation to reduce their use. All that could be agreed in respect of the GATT's provisions on export subsidies on primary products were changes in GATT rules which it was hoped would limit their effects. The changes involved the addition of new trade impact standards and the clarification of key concepts.

The previous GATT Article XVI:3 rule that agricultural export subsidies could not be used to gain 'more than an equitable share of world trade' was extended by Article 10 of the Subsidies Code requiring that export subsidies not 'displace' another country's exports to 'a particular market' by offering supplies at 'prices materially below' those of other suppliers. The concept of market shares 'in a previous representative period' was made more precise by the addition of the statement that this 'shall normally be the three most recent calendar years'.

Though the negotiations on the trade impacts of subsidies were the most important feature of the Tokyo Round, the subsidies code which finally emerged was not effective in disciplining the use of domestic and export subsidies in agriculture. The major concession that the United States had to offer – acceptance of an injury test in its countervail actions – was not sufficient to persuade the EEC to accept strict international rules on its use of subsidies in general, and in agriculture in particular. Hufbauer aptly summarized the situation in these words: 'when GATT Article XVI was drafted in 1955, US farm interests were too powerful to be disciplined: when code Article 10 was drafted in 1979, European farm interests were too powerful' (Hufbauer, 1983, p. 338).

Standards

Work which had started in the preparatory phase of the MTN was brought to a successful conclusion in an agreement on technical barriers to trade, commonly referred to as the Standards Code (BISD 26S/8–32). The substantive and procedural obligations of the agreement are fully applicable to trade in agricultural products.

Article XX(b) of the General Agreement allowed 'measures', including restrictions on imports, that were 'necessary to protect human, animal or plant life or health', subject to the constraints that such measures must generally be applied in a nondiscriminatory manner and not be used as a 'disguised restriction on international trade'. The Standards Code went further to ensure that the trade impacts of national standards-related activities – regulation, testing, and certification systems – would be minimized.

Signatories rededicated themselves not to use standards deliberately as concealed protection. They pledged not to adopt standards or procedures that were unnecessary obstacles to trade, where 'unnecessary' meant requirements that exceed what is required to protect legitimate public interests. It was agreed that national authorities would use international standards where these were available and appropriate. Advance public notice would be given and public hearings held when national standards were to be adopted or changed. Countries were encouraged to establish a single 'enquiry point' where foreigners could obtain information on national regulations, standards and certification procedures. Federal states were committed to using their best endeavours to ensure that these requirements would be observed by subordinate levels of government. For the first time, disputes in this area were brought into a specific multilateral forum within the GATT framework.

The Committee on Technical Barriers to Trade established to supervise the implementation of the agreement was empowered to establish panels of impartial experts to resolve disputes in this area, and to pass judgement over the remedial steps that should be taken.

This was an excellent start on the multilateral regulation of a set of non-tariff measures that were becoming increasingly significant barriers to trade. However, it was soon apparent that the application of the code to the field of agricultural and food products was not entirely satisfactory. Participation was low: only 37 countries signified their adherence to the code and this number did not include some important agricultural exporters. Progress in harmonizing international standards was slow, even where these existed. The role of international organizations was not spelled out. Countries were reluctant to accept the equivalence to their own of foreign testing methods and standards. The code required that, wherever appropriate, standards be specified in terms of product attributes, but some countries maintained standards for food and agricultural products based on production and processing methods which, they claimed, had an equivalent result. The jurisdiction of the code and the Committee over standards and regulations based on production and processing methods was challenged. The role of sound science in setting and resolving disputes over national standards and of risk assessment was not set out. More generally, the provisions of a generic code were not fine tuned to the specific sanitary and phytosanitary practices and standards of the food and agriculture sector. For these reasons, the regulation of standard-related matters in agriculture had to be carried further in the Uruguay Round.

Other Matters

Agreement was not reached on two other matters addressed in the MTN that were potentially important to agricultural trade.

Safeguards

A great deal of effort went into an attempt to reach an agreement on the multilateral safeguards system. This centred on clarifying and extending the provisions of the GATT Article XIX (Merciai, 1981, pp. 41–65, and Wolff, 1983, pp. 363–91). By the time negotiations closed, agreement had not yet been reached on the definition of the 'serious injury' standard which would permit a country to impose temporary import restrictions; the requirement that these be accompanied by 'adjustment assistance' measures; the multilateral control of 'voluntary export restraints' and like measures; and the 'special and differential treatment' to be accorded developing countries under any new regime. However, the negotiations finally failed on the issue of 'selectivity', that is, whether safeguard measures should be applied against all suppliers in a nondiscriminatory manner or selectively against those whose exports were judged to be 'disrupting' national markets.

A strengthened international safeguards system would have been valuable in the orderly conduct of international agricultural trade, particularly for trade in such commodity groups as fruits and vegetables where markets are chronically unstable and the temptation to take national safeguard measures is correspondingly great. A code on safeguards would also have been a useful adjunct to the International Arrangement Regarding Bovine Meat, and might have reduced demand for the counter-cyclical meat import laws which were introduced by the US and Canada in the early 1980s.

Consultative Mechanism

As early as 1972, the then Secretary-General of the GATT was putting the case for the establishment of a permanent body in the GATT which could address the particular problems of agricultural trade (White, 1972). The task of such a body would be the negotiation of ground rules for the conduct of national agricultural policies in the framework of the international trading system, and conflict avoidance and progressive policy harmonization through an annual review of policies measured against the consensual standards. This suggestion gathered wider support as the Tokyo Round proceeded, and 'there were various bilateral discussions about the formation of a high level group, relatively small in number of participants and relatively informal in its procedures, to discuss matters of common interest with respect to agricultural production, policy and trade.'[41] These discussions were not completed by the end of the negotiation, but the final communiqué pledged that the contracting parties would further develop active cooperation in the agricultural sector 'within an appropriate consultative framework.' Thereafter, the idea of creating a distinct body within the GATT for overseeing the development of national agricultural policies and international trade policies for agriculture was quietly dropped. At the 1980 session the CONTRACTING PARTIES formally decided that the Consultative Group of Eighteen – a high-level body which had been established by the GATT Council in 1975 – would also serve as a general forum on issues affecting agricultural trade. Plurilateral discussion of the trade effects of national farm policies and programmes then disappeared into the largely unpublished work of this Group and the even more informal 'Morges Group' of senior officials who met to discuss the continuing problems of agricultural trade policy. It was not until the Committee on Trade in Agriculture was established in 1982 that an effective mechanism for consultation on agriculture was created within the GATT (see Chapter 6).

5 CONCLUSIONS

The Tokyo Round as a whole had several notable achievements. Foremost among these was that governments were able to use the fact that they were engaged in trade negotiations to resist the demands of national groups for a protectionist response to inflationary recession at home and turbulent conditions in international markets for capital, resources, and goods. This was the first GATT round held in conditions of economic recession, rising unemployment and inflationary pressures in national economies. On the international scene, the confluence of crises in the international payments, food, and energy systems, the political assertiveness of the developing countries, and the emergence of new global competitors was without precedent. Economic nationalism and a protectionist firestorm could easily have been unleashed in the 1970s had multilateral trade negotiations not been joined.

The two most tangible results of the MTN were access improvements for manufactures and a useful beginning to the task of coming to grips with the trade effects of policies that operated primarily within national borders. These achievements have to be seen against a larger background, however. By any standards, the negotiations of the 1970s did not result in what the United States sought through its international economic diplomacy: a coherent set of jointly managed and refurbished international institutions in trade, money, and development, which would be the foundation of collective security and world economic order. The EEC and Japan did not have a sense of responsibility for the maintenance of international systems, nor did they share the United States' strategic vision. In the trade negotiations they were primarily concerned to establish their distinctive identities and defend existing national policies.

The MTN also fell short in some of the specific areas that were addressed. Negotiations failed to strengthen the international safeguards system. The developing countries obtained only niggardly improvements in access to rich-country markets. Some thought that at the end of the MTN the trading system itself was weakened in insidious ways. For instance, the codes fragmented the GATT dispute settlement system and extended the practice of 'conditionality' in GATT rights, and the Framework Agreement[42] which was concluded further eroded the most-favoured nation principle and encouraged noncompliance with basic GATT obligations by the developing countries.

The failure to deal with problems of trade in agriculture was also damaging to the GATT system and to international economic and political relations. There were several reasons for this failure. The EEC was not prepared to consider genuine agricultural trade reform in the 1970s. The CAP was only just getting into its stride. It was still the Community's only common sectoral policy. Its alteration in fundamental ways in response to complaints about its external effects was not politically feasible. It was easier to maintain internal support and to export the resultant surpluses than to lower prices or embrace production controls or income payments. EEC negotiators saw their task as protecting the Community's internal regulations from external constraints. In the early 1970s, Europe was moderately content with the CAP and determined – and confident enough – to defend it against outside interference.

But this places too much of the onus for the failure of the agricultural negotiations in the Tokyo Round on the European Community. The resolve of the other major participants to make the fundamental policy changes required for genuine agricultural trade reform may be questioned. The other countries of Western Europe had agricultural policies that were just as hostile to liberal agricultural trade as the EEC's CAP. Having yielded its market for wheat and feed grains to imports, and become a large and reliable importer of other food and agricultural products, Japan saw little need to take further steps to liberalize its agricultural trade. The US Congress's willingness to change American agricultural policies and trade arrangements was never tested. Canadian authorities were under strong pressure to increase protection for dairy, beef, and horticultural producers throughout the negotiations; they offered few concessions during the negotiations and intensified protection against imports of beef and horticultural products soon after they concluded. This left Australia and New Zealand to plead the case for a larger role for efficiency in the location of global food production, but they were small economies with little to offer and so had no bargaining power.

Though the US was capable of forcing agriculture onto the international agenda, by the time the negotiations opened, its initial focus on access and subsidized competition in the grains sector was made less urgent by the shortages, high prices and soaring exports values that developed in world grain markets. It is doubtful whether the EEC and Japan would have negotiated far-reaching changes in agricultural trade whatever the conditions of world markets, but the Ford Administration had no chance whatsoever in persuading these countries to negotiate on access and subsidization in the grains sector in the sellers' market that prevailed in its time. By the time the world food crisis was perceived to be the temporary result of a constellation of a set of unusual events, the Carter Administration was in office. In the negotiations on grains in the IWC, the US dissipated considerable political energy in pursuit of the stability, security and burden-sharing objectives that had been defined by the conditions which existed in an earlier period.

But even more importantly, by capitulating to Europe's demand for sectoral negotiations in agriculture and stating publicly that it would not seek fundamental changes in the CAP, the Carter Administration signalled that it wanted the MTN more than it wanted a far-reaching agreement on agricultural support and protection. Once this political reality was established, the requests and offers on access in Group Agriculture and the subsidy–countervail negotiations as they applied to agriculture were doomed to yield little.

It would require a major escalation of the external political costs and internal budgetary costs of operating contending farm programmes before the international community was ready to consider fundamental agricultural trade reform. This finally occurred in the Uruguay Round.

6 Markets, Policies, and Trade Rules in Crisis: 1979–86

As the Tokyo Round was concluded and the 1970s drew to a close, it was clear that the GATT system was still weak in the area of agricultural trade. Domestic agricultural policies were hardly constrained at all by international disciplines, and conditions in international trade were dominated by the influence of these national farm programmes. A large number of agricultural trade disputes were brought before the GATT, and even though most of these disputes were settled in a legal sense, these settlements did not have much of an effect on the way government policies impinged on trade flows. The situation had hardly improved at all since the inception of the GATT thirty years before.

Efforts to change this situation were undermined by world market developments in agricultural goods throughout much of the 1970s. These developments appeared to suggest that the major problems in agricultural trade were not over-subsidization and lack of market access, but global food scarcity and unreliable access to supplies. In the 1980s, economic conditions again changed fundamentally. World markets for agricultural products weakened, and international farm prices collapsed. As a result of these international market trends, and as a consequence of inflexible national policies, the financial and economic costs of farm policies rose sharply. This led to a severe farm policy crisis in most of the major industrialized countries. It was this crisis more than anything else that revived interest in the GATT as a forum for finding solutions to agricultural trade problems at the international level.

Action proceeded on two separate but parallel fronts. In the OECD, ministers of agriculture mandated a study of the effects, both domestic and international, of existing policies, and of the possible ways to mitigate the costs of these policies. Simultaneously, as part of the preparations for a new GATT round, a GATT Committee on Trade in Agriculture (CTA) was established. The Committee took stock of existing trade problems and discussed possible improvements to GATT rules for agriculture. As these discussions took place, market conditions continued

101

to worsen, and domestic farm policy costs mounted. Thus, when the final preparations were made for the Uruguay Round of GATT negotiations, the conceptual framework was in place and the economic conditions were such that it became both possible and urgent to tackle the crisis of agricultural markets, policies, and rules. In the end it became clear that agriculture had to be a major item in the agenda of the new round.[1] It is this crisis in the years immediately preceding the Uruguay Round negotiations on agriculture, and the discussions in the OECD and the CTA of the GATT, which form the subject of this chapter.

1   MARKETS IN CRISIS

The commodity price boom of the early 1970s had a profound impact on agricultural policies in the industrialized world. Agricultural capacity increased across the world in response to the high prices and increased market opportunities. Policy prices were increased, as world markets effectively removed the budgetary pressure. Acreage constraints were removed, in particular in the US, as farmers were encouraged to farm 'from hedgerow to hedgerow'. Increased costs, stemming in part from the rise in petroleum prices, added to the upward pressure on prices. The period of strong prices did not, however, last long. By 1976 the commodity price boom had run out of steam, and a more normal balance briefly returned to markets. World market prices of wheat, which had risen from their pre-1972 level of around $60 per metric ton to peak at $180 in 1973, dropped back to around $100 per metric ton by 1976. Other cereals also followed the same path, as did sugar and oil-seeds. Stocks rose, as governments rebuilt reserves and tried to maintain domestic prices. Budget pressures began to re-emerge, and consideration of policy reform returned uneasily to the agenda.

The downward drift in commodity prices was short-lived. If the price rises in the early 1970s were triggered by actual or imagined shortages, the late 1970s saw a climb in prices fuelled by demand. Over the decade, the volume of trade in agricultural goods had increased dramatically. World wheat trade, for example, doubled from 101 to 203 million metric tons between 1971 and 1980 (Hathaway, 1987). Demand for cereals and other products remained heavy, as developing countries entered the world market as major purchasers. To satisfy growing livestock industries and the burgeoning demand for meats required inputs of animal feed from abroad. Sales to the Soviet Union and other centrally planned economies also expanded rapidly, helped by a thaw in the Cold War. Some further tightening of supplies in the late 1970s also contributed to the rebound of world prices. By 1980 wheat prices had again reached their levels of the early 1970s, and fears of shortages returned afresh.

The reaction of major governments to this second price surge was predictable but unfortunate. Price supports were again increased in the EC and in the US, as the siren song of shortages lured politicians and analysts alike. In the US, the 1981 Farm Bill increased support prices and relaxed some of the constraints on agricultural production (Moyer and Josling, 1990). The production was forthcoming. US agricultural exports peaked in 1981, reaching a value of $44 billion. Firmer world prices for dairy products, due in part to the setting of minimum prices under the Tokyo Round International Dairy Arrangement, allowed the EC to enjoy a brief respite from expensive dairy subsidies in 1981 and 1982. Talk of policy reform was relegated to the academic literature and to the emerging consumer advocate groups.

The slide in world agricultural commodity prices from 1981 to 1986 was therefore perhaps the inevitable return of the chickens to roost. International demand was strangled by a combination of the debt crisis, a financial hangover from the borrowing binge of the 1970s, and the sharp recession of 1981–2, triggered by a second rise in oil prices. The effect was felt particularly strongly in the US, where the strength of the dollar, which rose by about 40 per cent in relation to other major currencies from 1981 to 1985, priced exports out of foreign markets and encouraged imports. In addition, repoliticization of US–Soviet trade in the wake of the invasion of Afghanistan reduced the market for US cereals. US wheat exports fell from 48 in 1981–2 to 25 million metric tons in 1985–6, and coarse grain exports plummeted from 71 to 36 million metric tons over the same period (Hathaway, 1987). The US government had to acquire substantial stocks, and devised an ingenious land-retirement programme which gave farmers grain from storage in exchange for restricting acreage (the Payment in Kind, or PIK programme). In total, 78 million acres were withheld from production in 1983, and nature contributed a sharp drought that year to reduce corn yields. Prices firmed briefly and then resumed their decline. Programme costs in the US rose sixfold between 1982 and 1986, peaking at $25 billion in that year.

The strong dollar should have helped the EC to contain programme costs over the first half of the 1980s. It is a sad reflection on the inability of Community politicians to constrain the CAP that the budget costs rose even as the dollar strengthened. Between 1981 and 1986, spending on the major CAP market support programmes doubled from 11 to 22 billion ECU.[2]

As if to emphasize the links between agricultural prosperity and financial markets, the high interest rates in the first part of the 1980s caused considerable problems for both farmers and countries. Those farmers who had borrowed heavily in the boom periods of the previous decade found that they had to repay the debt from depressed earnings. Real estate prices slumped, bankruptcies increased, and the threat of widespread rural depression became real. Countries also suffered from high rates of interest on debt acquired in the 1970s, and began to contract domestic spending as well as imports. When escalating farm programme costs met constrained budgets, the first serious thought was given to radical reform of farm policies.

The EC faced a particularly difficult problem in the dairy market. Attempts to keep up the income of millions of dairy farmers through the support of the price of a handful of traded dairy products, such as butter, cheese, and milk powder, broke down under the weight of depressed world prices for these products and expanded production by large modern dairy farms. Rather than reduce support prices, the EC opted for quotas to control production. These quotas helped in the short run to curb surpluses but in the long run inhibited the development of a low-cost, competitive dairy sector.

The US reaction to the policy crisis was only a little more far-sighted. A Reagan Administration draft of the 1985 Farm Bill proposed lower price supports and a phasing out of the deficiency payments system. Financial distress in the agricultural heartland dampened law-makers' enthusiasm for radical solutions, and they settled for a programme which expanded exports, withdrew marginal land from production, and lowered support prices over a period of five years. In addition, the Bill partially 'decoupled' deficiency programmes from current output by freezing acreage bases and yields for each farm. The new farm policy therefore both began the process of policy reform and gave the US an additional weapon for use in the coming trade war. At the same time the US decided that a subsidy war might hold more promise than reliance on law in the GATT. A programme of export subsidies was introduced, targeted to markets the US believed it had lost to the EC as a result of EC export subsidies.[3] This Export Enhancement Program (EEP) was thought to both warn the EC that its export subsidization would not remain unchallenged and to drive the EC to the negotiating table. Even though the EC would have preferred to respond not too aggressively to this subsidy war, the only way it could dispose of its continuing and growing surplus production was by matching US export subsidies. At that stage, both parties, as well as other countries, had essentially given up on pursuing complaints about agricultural export subsidies through the GATT legal mechanism.

The sense of crisis in agricultural policies was widespread in the mid-1980s. It was the subject of considerable academic discussion and quantitative analysis. The political response was to develop both domestic and international approaches to the problem.[4] Thus a number of institutions became involved in the search for more satisfactory policies. Two such institutional responses, those that took place in the OECD and in the CTA of the GATT, proved to be important in the preparation of the Uruguay Round agenda. With a fine sense of comparative advantage, the OECD addressed the issue of domestic policy reform, the reinstrumentation of farm programmes in the developed countries to achieve a more market-oriented agricultural sector, and the CTA of the GATT took on the question of the multilateral rule changes which would lead these policies to avoid serious interference with the functioning of international agricultural markets.

2 FARM POLICIES IN CRISIS: THE OECD TRADE MANDATE

The Organization for Economic Cooperation and Development (OECD) was the institution chosen by the major developed countries, some with more enthusiasm than others, to tackle the international dimension of the crisis in farm policies. OECD members embarked in 1982 on an analytical effort on agricultural policies and trade which in the end proved an important element in the preparation of the political and intellectual ground for the Uruguay Round negotiations on agriculture. This OECD study, in response to the Ministerial Council Mandate on Agricultural Trade (known in short as the Ministerial Trade Mandate, MTM) was a remarkable departure from earlier work on agriculture in the OECD. In the past, the OECD, as befitted a body designed for policy consultation and analysis rather than as a negotiating forum, had shied away from pointing a finger at individual member countries on contentious agricultural issues. The result was a tendency to engage in somewhat anodyne descriptive work and to avoid analytical studies which could have produced quantitative information on the policies of individual countries and their effects on agricultural trade. In its work on agriculture, the OECD secretariat had never been encouraged to use analytical methodologies such as estimating rates of protection, presumably because some countries feared that they could show how 'guilty' their agricultural policies were.

A good example of the earlier approach was an OECD study on agricultural trade which had been produced just before the MTM. This report on *Problems of Agricultural Trade* (OECD, 1982a) collected information on trends in agricultural trade and discussed at length the general relationships between various policies and trade. It also pointed out 'that agricultural trade is distorted by interventions of all kinds by practically all countries'. However, no attempt was made to quantify the size of distortions in any particular country. The farthest that the report went in this direction was to reproduce one table on effective rates of assistance in Australia from the Australian Industries Assistance Commission (1980) as 'an example of the results which could be obtained by these types of calculations'. Also, the study did not analyse in any way the effects which individual countries' policies had on trade. It rather said that

> in order to assess the extent of the distorting effects of these interventionist and usually protectionist policies, it would have been necessary to estimate the situation – in terms of production, consumption, trade and prices – which would have resulted in the world in the absence of these measures. This is, of course, extremely difficult and no attempt has been made here to make these assessments. Research bodies should, however, be encouraged to proceed in this direction.

Immediately after that report, however, the OECD itself proceeded in exactly that direction.

At its meeting in May 1982, the OECD Council at ministerial level endorsed the conclusions of the study on *Problems of Agricultural Trade* and, at the insistence of countries such as the United States, Australia, and New Zealand, and in spite of some hesitation on the side of the EC, invited the Committees for Agriculture and for Trade to undertake jointly a study contributing 'to progress in strengthening co-operation on agricultural trade issues and to the development of practical multilateral and other solutions' (OECD, 1982b). The study was to have three elements:

(i)   an analysis of the approaches and methods for a balanced and gradual reduction of protection for agriculture, and the fuller integration of agriculture within the open multilateral trading system, while taking into account the specific characteristics and role of agriculture . . .;

(ii) an examination of relevant national policies and measures which have a significant impact on agricultural trade with the aim of assisting policy-makers in the preparation and implementation of agricultural policies;

(iii) an analysis of the most appropriate methods for improving the functioning of the world agricultural market. . . .

Work on the MTM was undertaken mainly in the OECD Directorate for Food, Agriculture and Fisheries, under its director Albert Simantov, with help from the Trade Directorate under the auspices of the Joint Working Party of the Committee for Agriculture and the Trade Committee, chaired at the time by George E. Rossmiller from the Foreign Agricultural Service of the US Department of Agriculture. After much deliberation within the OECD Directorates, several meetings of the Joint Working Party, and consultation with sundry academic economists, the decision was taken in 1983 to base the MTM primarily on two analytical methodologies. The extent of assistance provided to agriculture in individual countries was to be analysed using the concepts of Producer Subsidy Equivalent (PSE) and Consumer Subsidy Equivalent (CSE) as developed by Josling for the FAO a decade earlier (FAO, 1973, 1975). Second, the quantitative impact of policies on trade was to be estimated with the help of a multi-commodity trade model developed within the OECD Agriculture Directorate. PSE and CSE as measured for a number of core agricultural products and for Australia, Austria, Canada, the EC, New Zealand, Japan, and the United States (and, on a more aggregate basis, for the Nordic and Mediterranean OECD countries) were first entered into that model at their base period (1979–81) values and then at a level reduced by 10 per cent. World price impact and effects on volumes traded for this reduction in the level of assistance were estimated.

A significant part of the staff and resources of the OECD Agriculture Directorate was invested in this rather ambitious analytical exercise over the period 1983 to 1985, and discussions on it in the Joint Working Party and the Agriculture and Trade Committees were not always harmonious. However, the results appeared to be worth the effort. The directly tangible outcome of the MTM was a series of OECD publications which appeared in 1987, shortly after the Uruguay Round had been launched. A synthesis report on *National Policies and Agricultural Trade* (OECD, 1987) presented the overall results. It was

accompanied by seven volumes, one each for the six individual countries and one for the EC, giving details of the calculations.[5]

The synthesis report presented results of the PSE and CSE estimates
and concluded that 'all Member countries support their agriculture and
intervene in the functioning of the market, thus preventing market forces
from operating effectively, particularly by support policies which are
generally accompanied by border measures: however the level and
modalities of assistance vary widely between countries'. It thus established the fact that all OECD countries were sinners in agriculture,
though to varying degrees. The PSE and CSE figures presented in that
report were the first 'official' estimates of the level of support provided to agriculture in the countries concerned. The fact that an institution which represented governments had made such figures public
was in itself a great achievement. The study created a new dimension
of transparency in agricultural policies at the international level, showing
that between 15.5 per cent (New Zealand) and 59.4 per cent (Japan) of
the value of agricultural production resulted from government policies
rather than from markets. It also established the fact that governments
could agree on a methodology for producing such politically sensitive
figures.

The results of the trade modelling exercise under the MTM were
noteworthy for a number of reasons. The model developed by the staff
of the Agriculture Directorate was methodologically sound and compared favourably with the models available at the time in academic
circles. In particular, it was strong in terms of the cross-commodity
linkages, a feature which was important for the policy message it was
supposed to generate. The MTM study was only used to estimate the
effects of only 10 per cent reduction in levels of support, because that
was considered to be in line with the 'gradual' reduction in support
called for by ministers. The 'balanced' nature of reductions was interpreted, on the insistence of some countries, in the form of five different scenarios, including not only a straight proportional reduction of
PSE across the board for all countries and commodities, but also the
same overall support reduction allocated in different ways to individual
countries and products. In terms of political courage, presentation of
the model results was itself remarkable. These were not confined to
the world market price effect of a multilateral support reduction in all
OECD countries: they also demonstrated what the contribution of each
individual country was in that overall effect. Thus it was shown, for
example, that with a multilateral 10 per cent reduction in assistance
the world market price for dairy products would increase by 4.4 per

cent, of which 2.8 percentage points resulted from support reduction in the EC alone (OECD, 1987, p. 32). Thus the identification of particularly egregious policy distortions, which had been so carefully avoided earlier, was now done with precision.

The results of the model analysis held no surprises (except perhaps for the fact that the OECD model was the first to predict that world prices for some commodities could actually go down with liberalization, a result of both the cross-commodity features of the model and the inclusion of supply control policies and their relaxation in the support reduction scenario). But they were important in that they were produced by an intergovernmental institution. Moreover, the estimates were timely and easily understood by policy-makers. This helped to establish a number of insights among governments which academics had not sufficiently managed to get across to that audience. One message was to emphasize the link between domestic and trade policies. It was no longer possible for countries to argue that their policies were of no concern to others. A second lesson was that world prices (of most products) might increase with multilateral support reduction such that liberalization was not quite as politically dangerous as it might appear. Hence a gradual and balanced reduction in support in which all countries participated would have a less dramatic effect on domestic markets and prices than often feared. A third important message was that market links among products were extremely important, and that it was therefore essential that all products be included in a liberalization exercise. As an additional benefit, the measurement of PSE contributed to the fuller understanding of the way agricultural policies worked, the magnitude of the transfers involved, and where the true costs of those policies fell.

In addition to the information of value in trade discussions, the reports produced under the MTM also had a wealth of other quantitative and qualitative information on agricultural policies in OECD countries and on agricultural trade issues. The synthesis report, for example, gave numbers on public expenditure on agricultural policies in OECD countries, and pointed out that it had significantly grown between 1979 and 1985. It also made it clear, though, that public expenditure was only about one half of all transfers to farmers in OECD countries, the other half coming from consumers. The total sum of transfers in OECD countries, put at about 100 billion ECU per year on average in the 1979–81 period and much higher in subsequent years, caught media attention. The country reports had useful and detailed information on agricultural policies in the countries concerned. While PSE/CSE estimates and the trade model appropriately responded to the first two items of the Ministerial Mandate – to analyse approaches to a balanced reduction of protection and to identify policies with a significant impact on trade – the final report on the exercise did not have much to say about the third item, i.e. the most appropriate methods for improving the functioning of world markets. That may have been intentional, given that the OECD was not intended as a negotiating forum. Moreover, by the time the report was published the GATT negotiations had already been launched.

Overall the MTM exercise was a definite success, and an important element in the preparations for the Uruguay Round negotiations in agriculture. After that exercise, no OECD country, with the possible exception of New Zealand, could any longer claim that it did not contribute to the disarray in world agriculture. The traditional argument that countries needed to protect their domestic markets against unfair policies in other countries had lost much of its political appeal. Journalists now had, from an official source rather than from obscure academic research, all the numbers they needed to inform the general public about the large transfers involved in agricultural policies and about the damage agricultural protectionism did to international trade. Above all, it had been made clear, in an analytical exercise in which government representatives were directly involved, that liberalization could improve the state of affairs in agricultural trade. The 'official' establishment of the PSE as an all-embracing common denominator for the widely varying forms of agricultural protection and support had brought home the message that it was the totality of agricultural policies, both at the domestic level and at the borders, which needed to be addressed in multilateral negotiations. Moreover, the fact that the PSE measure had now been widely publicized among governments made it possible to consider the use of that measure as a tool for negotiations and commitments in the Uruguay Round. The actual use which was made of an aggregate measurement of support in the final Agreement on Agriculture fell short of some early expectations. However, as will be discussed in the following chapter, the emphasis placed on such an aggregate indicator of support and protection in the early years of the Uruguay Round helped to focus the mind of negotiators on the need to adopt an all-embracing and across-the-board approach, rather than slipping back to the traditional approach of engaging in requests and offers on specific border measures for particular products in individual countries.

3 TRADE RULES IN CRISIS: THE GATT COMMITTEE ON TRADE IN AGRICULTURE

Establishment of the GATT Committee on Trade in Agriculture

In 1981, the Consultative Group of Eighteen (see Chapter 5) recommended that the next annual GATT Session should be held at ministerial level. It had been eight years since the previous GATT ministerial meeting had been called, to start the Tokyo Round. The new Ministerial Session was scheduled for November 1982 in Geneva, and was supposed to 'examine the functioning of the multilateral trading system, and to reinforce the common efforts of the contracting parties to support and improve the system for the benefit of all nations.' (GATT, 1982a, p. 27). During the preparations for the session it became obvious that this was not going to be a harmonious event. Discussions during the meeting were extremely tense, and on a number of important issues no consensus was reached. Though it was clear from the beginning that agricultural issues would be important during the ministerial session, they proved to be even more divisive than expected. Indeed, agriculture emerged as a subject of major controversy. A number of reservations expressed by several countries with regard to the concluding Declaration originated from their dissent on agricultural issues.

In the Ministerial Declaration, the contracting parties outlined a 'work programme and priorities for the 1980s' in which they undertook, among others things, to bring agriculture more fully into the multilateral trading system by improving the effectiveness of GATT rules, provisions and disciplines and through their common interpretation; to seek to improve the terms of access to markets; and to bring export competition under greater discipline' (BISD 29/11–12). However, the most important concrete result in agriculture of the 1982 Ministerial Session was the establishment of a new GATT body to look into agricultural matters, the Committee on Trade in Agriculture (CTA). In historical perspective the CTA can be seen as establishing a link between the Tokyo Round and the Uruguay Round in the GATT's dealings with agriculture.

Apart from the few agricultural commitments which had been negotiated bilaterally, and the two largely inconsequential arrangements on dairy products and beef, the Tokyo Round had not resulted in any significant change in the conditions governing agriculture in the GATT. No new institution (apart from the dairy and beef councils) to deal with agriculture had been created. A consultative council on agricultural matters in the GATT as proposed during the Tokyo Round had not in fact been established, and the informal Morges Group of some of the major agricultural trading countries had neither become an official GATT body, nor had it managed to achieve anything substantial (see Chapter 5). Agricultural issues were discussed with some intensity in the Consultative Group of Eighteen, but most results of these deliberations were not made public.[6] Hence the CTA was the first post-Tokyo Round institution in the GATT to deal explicitly with the problems in agricultural trade. Its significance for the later Uruguay Round lies partly in the fact that the 1982 GATT Ministerial Session can be seen as the origin of that new round of multilateral negotiations. Also, the chairman of the CTA, Aart de Zeeuw from the Netherlands, later became the chairman of the Uruguay Round Negotiating Group on Agriculture. Moreover, the CTA foreshadowed the Uruguay Round negotiations on agriculture because it discussed the various options for how improvement in GATT rules for agriculture could be negotiated.

The mandate of the CTA was to examine 'all measures affecting trade, market access and competition and supply in agricultural products, including subsidies and other forms of assistance . . .', 'in the light of the objectives, principles and relevant provisions of the General Agreement and also taking into account the effects of national agricultural policies, with the purpose of making appropriate recommendations' within a two-year period. The CTA was expected to take 'full account . . . of specific characteristics and problems in agriculture, of the scope for improving the operation of GATT rules, provisions and disciplines and agreed interpretations of its provisions' (BISD 29S/16–17).

The fact that the need was seen in the early 1980s to reconsider GATT rules on agriculture seriously, and that the CTA was mandated to take another look at these rules, reflected in part the crisis on international markets and in domestic agricultural policies in many countries. It also demonstrated that these rules had simply proved unsatisfactory. Many agricultural policies effectively escaped GATT discipline, in large measure due to the nature of the respective GATT rules. It is therefore useful to take a look at these GATT rules for agriculture in the next section before describing how they were discussed in the CTA.

Problems with GATT Rules on Agriculture

It is sometimes said that until the Uruguay Round agriculture had remained largely outside the GATT. As should be clear from the preceding chapters, this was certainly not the case. Agriculture did play a role in the various rounds of GATT negotiations, and in the day-to-day business of the GATT. Also it would be wrong to assume that the rules of the GATT did not apply to agriculture. On the contrary, all 38 Articles of the General Agreement have always applied to agriculture, as they apply to trade in manufactures.[7] Agriculture has not had a general exemption from GATT disciplines: instead a few special exceptions for agriculture were built into the Agreement. Two of these specific provisions for agriculture in the GATT are particularly important.[8] One applies to rules for imports (Article XI:2(c)), the other to rules for exports (Article XVI:3). In the remainder of this chapter the importance of these import and export rules will be discussed, along with the rules regarding domestic subsidies.[9] The conclusions drawn from relevant GATT disputes are partly based on the excellent documentation and analysis of GATT dispute settlement by Hudec (1993).

The GATT and Import Measures in Agriculture

In general, all provisions of the GATT regarding the treatment of imports apply to agricultural products as they do to industrial products. In principle this means that the only border measures to be used should be ordinary customs duties, preferably tariffs bound in Schedules annexed to the Agreement. There was however one provision in the GATT 1947[10] which made a distinction between agriculture and industry when it came to the legality of quantitative restrictions. This was the infamous provision of Article XI:2(c), regarding quantitative restrictions in agriculture. The historical origin and the text of this provision have been related in Chapter 1.

The agricultural clause of Article XI:2(c) was an exception to the general ban on quantitative restrictions contained in Article XI:1. It allowed import restrictions on agricultural products if they were necessary to enforce certain forms of domestic market management. The most important domestic measure potentially qualifying for an exception from the prohibition of import quotas was direct domestic supply control (subparagraph (i) of Article XI:2(c)).[11] However, a number of conditions had to be met before this provision could be invoked. These conditions were always contentious, and a significant number of disputes have been brought before the GATT regarding their interpretation.[12] In general, the requirements to be met have been interpreted rather strictly by GATT panels, and panels have therefore effectively constrained the legal use of this agricultural exception.[13] The difficulties encountered in GATT disputes about this provision were mainly of a legal nature. However, there are also some fundamental economic problems with Article XI:2(c).[14]

As far as the legal issues are concerned, complaints related mainly to the proper interpretation of the criteria required to invoke Article XI:2(c). Four questions were of particular importance in GATT disputes: what is an effective restriction of domestic supply; when is an import restriction necessary to enforce domestic supply control; to which products can import restrictions be applied; and what is the minimum volume of imports that must be allowed in?

The 'effective restriction' criterion was interpreted rather strictly by panels. According to their rulings, essentially only quotas for individual producers, with sufficiently harsh sanctions for producing beyond quota volumes, could meet this criterion. Typical examples of domestic restrictions accepted by panels were dairy quotas in Canada and acreage quotas for legumes and groundnuts in Japan.[15] However, the mere existence of such quota schemes was not considered *prima facie* sufficient, because it was not, in the view of panels, necessarily clear that they kept output below the level which it would have attained in the absence of the supply restriction, as intended by the drafters of the General Agreement.[16] Panels were not satisfied with less stringent forms of supply management, such as policies that provided producers with the option of withholding their output from the market with government support if the market price dropped below a given level,[17] or with sporadic government measures such as the occasional grubbing up of apple trees.[18]

Panels have also been strict in determining whether import restrictions were 'necessary' to the enforcement of domestic supply controls. They have found them not necessary, and hence not justified by Article XI:2(c), if they were applied in seasons in which the domestic supply of a perishable product (such as tomatoes) did not come on the market and in which, therefore, domestic supply was not restricted.[19] They have also denied the necessity of restricting imports which were so small relative to the domestic market volume that they could not possibly undermine the operation of the domestic supply control.[20] Moreover, import restrictions have been found to be unnecessary if they applied only to a given processed product, while imports of the

same product at earlier stages of processing were not restricted.[21]
The proper definition of the characteristics of products to which import restrictions might be applied has caused some headaches. Can imports of tomato concentrate, for instance, be restricted if the domestic supply of tomatoes is controlled? Do quotas on domestic milk production justify quantitative restrictions on the importation of ice cream and yoghurt? Obviously, product relations can come in a vertical dimension (different stages of processing, i.e. the issue of agricultural product 'in any form') and a horizontal dimension (different varieties of a product, i.e. the issue of 'like product'), and the drafters of this provision appear to have thought of both these dimensions.[22] In agriculture, the vertical dimension is particularly relevant as unprocessed agricultural products sometimes cannot be traded internationally.[23]

Panels have argued at length about the controversial issue of the proper product definition, with much legal sophistication. One of the many arguments considered was that Article XI:2(c) was not supposed to allow protection of the domestic processing industry, but then it was not always clear how one should determine whether an import restriction on a processed product provides protection to the processing industry if its raw material benefits from price support as provided under the umbrella of domestic supply control.[24] Overall, panels have tended to interpret the criteria of 'like product' and 'in any form' narrowly too, probably so as not to open a flood gate to the use of Article XI:2(c). However, one can still find some ambiguity and vagueness in the ways the issue of proper product definition has been 'resolved'. This is not surprising given the (possibly unavoidable) lack of precision of the relevant provisions in the General Agreement (Hartwig, 1992, pp. 120ff.). In the end, the difficulties of agreeing on a proper product definition have greatly added to the uncertainties surrounding the applicability of Article XI:2(c).

Interestingly enough, the one provision in Article XI:2(c) which has not figured very prominently in GATT disputes is the minimum quantity of imports to be allowed in. If a country restricts imports under Article XI:2(c)(i), the last subparagraph of this provision requires that it must not 'reduce the total of imports relative to the total of domestic production, as compared with the proportion which might reasonably be expected to rule between the two in the absence of restrictions', paying regard 'to the proportion prevailing during a previous representative period and to any special factors' affecting trade. That this issue has not played a larger role in panel deliberations does not reflect its irrelevance, but rather that panels have in most cases found other reasons for rejecting the invocation of Article XI:2(c). Hence panels did not feel required to consider the appropriate size of an import quota under Article XI:2(c), much to their relief, one may assume, given the difficulties of finding a practicable interpretation of this clause in the General Agreement.

In the typical case of domestic supply controls, which may have been in existence for a long time, and where all sorts of factors, including various government policies in addition to supply control, may have affected the level of domestic production, it is 'virtually impossible to objectively determine what the level of production would have been in the absence of restrictions' as one panel quite rightly noted.[25] The same panel also found 'that the burden of providing the evidence that all the requirements of Article XI:2(c)(i), including the proportionality requirement, had been met must remain fully with the contracting party invoking that provision'. These two findings taken together came close to suggesting that Article XI:2(c) could not really ever have been successfully invoked in practice. This was probably a desirable situation: after all, this GATT exception for agriculture did not really make good economic sense. Not only did it run counter to the general GATT spirit of a liberal trading system based on a tariffs-only philosophy, its internal economic logic was not really convincing.

Even if it were not for this last-mentioned logical trap, the totality of the requirements to be met before import restrictions could be imposed under Article XI:2(c) was so demanding, and has generally been interpreted so restrictively by panels, that this exception for agriculture in the GATT cannot be said to have opened up a loophole for the use of non-tariff barriers in agricultural trade. Out of the 16 GATT disputes which related to Article XI:2(c), there was not one single case in which the import restriction was found to be consistent with this provision.[26] Other cases where countries have maintained import quotas ostensibly under Article XI:2(c) were not tested in GATT disputes: their consistency with this provision never needed to be demonstrated. Article XI:2(c) has therefore had more of a symbolic than a practical role in agricultural trade policy.

What then was the reason why non-tariff import barriers abounded in agriculture before the Uruguay Round? The many non-tariff measures in agriculture that did not rely on Article XI:2(c) came mainly under three categories. First, in agriculture much use has been made of grandfather clauses and protocols of accession to maintain non-tariff measures.[27] Second, in one important case a waiver had been obtained which essentially allowed the use of quantitative restrictions even though the requirements of Article XI:2(c) were not met. This was the famous US waiver whose history, and relation to Article XI:2(c), is described in Chapter 2. Even though the impact of the US waiver may have been limited in purely quantitative terms, the damage it did to the GATT trading system was significant.

Third, in addition to country-specific derogations and exemptions, some countries relied on the use of so-called grey-area import measures. Trade measures have generally been described as falling in the 'grey area' of the GATT when they were of a nature not explicitly regulated in the GATT. At closer inspection, grey-area measures come in two different forms. On the one hand there are certain measures that tend to be used to restrict trade in products for which the importing country has a bound tariff. In this group, voluntary export restraints and orderly marketing agreements are particularly important. A typical case in agriculture was the voluntary export restraint agreement concluded between the EC and Thailand in 1982.[28] A limited number of such voluntary restraint agreements existed in agriculture, but it appears that this particular trade policy instrument was no more, and probably less, common in agriculture than in industry. As long as the importing country did not complain, there was nothing in the General Agreement which prevented an exporting country from restraining its exports 'voluntarily'.[29] On the other hand, importing countries often engaged in 'armtwisting' by threatening to use, or actually using, discriminatory import restrictions, whether GATT-legal or not, until the exporting country 'voluntarily' restrained its exports.

Much more important in agriculture was a second form of grey-area measures, where the importing country did not have a tariff binding or, in some cases, where full use was not made of the bound tariff rate. Variable levies, as extensively used by the European Community, minimum import price schemes and similar measures fall in that category. These measures are 'grey' in the sense that their conformity with the GATT has never been explicitly confirmed or denied.

The relevant GATT provision is that of Article XI:1, which bans 'prohibitions or restrictions other than duties, taxes or other charges, whether made effective through quotas or . . . other measures'. It is clear that this provision outlaws quantitative restrictions. Indeed, the heading of Article XI is 'General Elimination of Quantitative Restrictions'. It is less clear whether price-related NTM such as variable levies were outlawed as well. The explicit mention of quantitative restrictions only in the heading did not exclude its applicability to other forms of NTM, though it may have distracted attention from them. It could well be argued that Article XI:1 encompassed, and banned, all NTM which were not regulated in other GATT provisions (see Dam, 1970, p. 151). More specifically, variable levies could potentially be assumed to fall in the category of 'other measures' prohibited in Article XI:1. This could, in particular, appear to be the case if a levy is set at a level so high that it is prohibitive, in which case there is no difference in effect between that levy and an import prohibition.[30]

On the other hand it can be argued that a variable levy falls in the category of 'duties, taxes or other charges' permitted in Article XI:1. It is distinct from other tariffs 'only' in the sense that it is varied from time to time. As long as a tariff has not been bound for the product concerned, why should the tariff not be changed occasionally?[31] Moreover, even where a tariff binding exists, if a variable levy for the product concerned is set such that it never exceeds the bound tariff rate, why should the variability as such violate the GATT? Is a high but constant tariff more GATT-consistent than a lower but variable duty?

In the past, these questions relating to price-related grey-area measures have not been settled in the GATT. There was only one complaint in the early history of the CAP that has been filed against (among others) the variable levies of the EC.[32] However, the panel dealing with this complaint 'concluded it was "not appropriate" to rule on the legality of EC variable levies, due to the existence of unresolved disagreement after previous consideration by the CONTRACTING PARTIES' (Hudec, 1993, p. 446). In effect, the legality of price-related grey-area measures, and in particular variable levies, has not been decided in the GATT. Given the significance of these measures in agricultural trade prior to the Uruguay Round this may be suprising. The lack of legal GATT challenge to these measures certainly does not mean that all countries have considered them appropriate under the GATT. It rather reflects the feeling that the chances of constraining their use effectively through GATT proceedings were very small, not least because of the fundamental importance the EC had attached to them in designing its CAP. As long as the EC was not willing to see the CAP undermined in the GATT, there was little hope of coming to grips with these grey-area measures which constituted barriers to agricultural imports.

The GATT and Export Measures in Agriculture

With respect to export measures, there was also one infamous exception for agriculture in the GATT 1947. In industry, export subsidies were plainly prohibited (Article XVI:4). In agriculture, export subsidies were legal, as long as they were not used to gain 'more than an equitable share of world trade' in the product concerned (Article XVI:3). These rules on export subsidies, both the general ban and the agricultural exception, were not originally included in the General Agreement. They were added in 1955, in a period and through a process described in Chapter 2, where the relevant parts of Article XVI:3 are also cited. During the Tokyo Round, an attempt was made at strengthening GATT rules on subsidies, by means of the Subsidies Code. Article 10 of that code aims at clarifying and adding precision to the GATT rules on export subsidies in agriculture. The negotiations which led to that code have been described in Chapter 5.

The exception regarding export measures in the GATT 1947 triggered as many disputes as did the exception regarding import measures.[33] However, in all other regards the export exception was different from the import exception. Whereas panel rulings on Article XI:2(c) were unequivocal and effective, in the area of agricultural export subsidies they tended to be weak and permissive. Subsidizing countries were rarely found to violate their GATT obligations, and much use was made of the agricultural exception regarding export subsidies.

Two elements in Article XVI:3 appeared to be simple to interpret. First, panels had no problems finding that the policies complained about were indeed export subsidies, and the defendant countries did not contest these findings. This may be surprising, as Article XVI does not contain a definition of 'subsidy' nor of 'export subsidy'.[34] However, it may well reflect the fact that governments indeed used rather overt export subsidies in agriculture because they did not even feel the need to employ more refined forms of export assistance which could arguably have escaped the definition of an export subsidy. Second, no problems were encountered in GATT disputes with the definition of 'primary products'. This is somewhat worrying because of the ease with which some processed agricultural products were implicitly considered to be primary products. This issue is taken up below.

The real difficulty arising out of Article XVI:3 was with the concept of an 'equitable share of world export trade', including the need to take account 'of the shares . . . during a previous representative period, and any special factors which may have affected or may be affecting . . . trade'. The first GATT panel ruling on a dispute over agricultural export subsidies, in 1958, did not encounter major difficulties with this concept. Looking at world trade statistics, which indicated that the world market share of the defendant country had about doubled over the period considered, the panel found that exports of the defendant country had risen 'very substantially' and had 'remained considerably' higher than in the past. On that basis the panel ruled, in a rather straight-forward manner, that the export subsidies concerned had contributed to the increase in exports, that the resulting market share was more than equitable, and that exports of the complaining country had been displaced and therefore that its interests had been damaged. The CON-TRACTING PARTIES adopted the panel report, and the two countries agreed on how to settle the conflict.[35] However, this was the only GATT dis-pute over agricultural export subsidies which was 'successful' in the sense that the defendant country was found to violate Article XVI:3 or the corresponding provisions in the Subsidies Code. Most of the later cases were not even pursued to the point that a panel ruling was reached, and the only three other disputes over the GATT Article XVI:3 or Article 10 of the Subsidies Code where a ruling was made ended with a verdict of 'not guilty' for the subsidizing party (always the EC).[36]

The four problems that panels had with the equitable share rule re-lated mainly to assessment of the relevant market in which to measure the 'share'; the relevant quantities of exports to consider; the appro-priate reference period to use; and the causality between the export subsidy and the market share (Hartwig, 1992, pp. 179–88).

The relevant market for assessing the 'share in world export trade' could be either the aggregate world market or markets in individual importing countries. In the one early 'successful' case the panel took a significant increase in the exporting country's share in total world trade of the product concerned to be evidence of its having exceeded its 'equitable share'. The panel looked at individual destinations only when it came to assess the damage done to the complaining country. In the two cases which followed, relating to EC sugar exports, the panels did not find an increase in EC exports (which nearly doubled the world market share of the EC) sufficient to conclude that the EC had exceeded its 'equitable share'. The panels instead looked at differ-ent parts of the world market to see whether EC exports had displaced exports of the complaining countries.

The concept of displacement had been agreed as one possible inter-pretation of 'more than an equitable share' in the Subsidies Code (Article 10:2(a)) during the Tokyo Round in an attempt to provide this GATT rule with more teeth. However, as used by panels the concept of 'displacement' made Article XVI:3 even more toothless. In the two sugar disputes the panels found a few markets where exports from both the EC and the complainant countries had decreased, and they took this as a reason to conclude that displacement could not be clearly established.[37] The last panel to make a ruling in an agricultural export subsidy dispute, dealing with a complaint of the US about EC exports of wheat flour, had even more difficulty in finding displacement on individual markets, even though the case appeared rather clear. The US could demonstrate that it had lost trade to the EC on 17 individual markets, but the panel, though not questioning the accuracy of the data presented, still found that the markets concerned 'had changed considerably in size and nature over the period, and the changes in market shares were such that cases of displacement . . . [were] not evident'.[38]

The issue of the relevant export quantities to be taken into account was debated mainly in one case, but the difficulty encountered was symptomatic of much of agricultural trade. In a US complaint about EC exports of wheat flour, the EC wanted to see US concessional exports (mainly food aid) included in the size of 'world export trade', while the US opposed that view. After a long debate among the parties, which raised arguments worthy of a theological disputation, the panel decided to include concessional shipments in the analysis. However, an explicit decision was never taken by the CONTRACTING PARTIES as to whether that was the appropriate approach.[39]

Not surprisingly there was also much disagreement over which 'previous representative period' should be chosen in each individual dispute. Agricultural markets are known to be notoriously volatile, and trade flows can change rapidly. In that sense there is rarely any individual period which is 'representative'. Moreover, export subsidies have been granted in agriculture for such a long time that a period representative of the 'true' market shares countries would have in the absence of export subsidization is not really available.[40] In one case, the panel looked back as far as 24 years to find such a period.[41] Later panels experimented with different reference periods. The Subsidies Code had tried to create more clarity by stipulating that '"a previous representative period" shall normally be the three most recent calendar years in which normal market conditions existed'. However, panels still found it difficult to recognize 'normal market conditions'.

The greatest difficulty panels had, however, was to establish causality between export subsidies and expansion of market share. After the first and only successful case, each of the later panels was overwhelmed by the multiplicity of 'special factors' at work in the markets concerned. Because of these many 'special factors' and their complexity, panels felt unable to find that the rising market shares of the exporting country were mainly due to export subsidies. The most impressive example was the panel report over agricultural export subsidies, dealing with EC exports of wheat flour. The panel said 'it was evident . . . that the EC share of world export . . . had increased considerably over the period under consideration when application of EC export subsidies was the general practice, while the share of the US and other suppliers had decreased' (GATT, 1983, p. 33). The panel also 'found it anomalous . . . that the EC which without the application of export subsidies would generally not be in a position to export substantial quantities of wheat flour, had over time increased its share of the world market to become by far the largest exporter'. However, the EC cited a long list of special conditions, including the close links between markets for wheat and flour; political relationships between importing and exporting countries; the influence of US embargoes; the existence of regular shipping lines between the EC and the importing countries; cultural and linguistic ties; preference for EC flour qualities; and impacts of concessional sales.[42] Given all of these conditions, 'the panel found . . . that it was unable to conclude as to whether the increased share has resulted in the EC "having more than an equitable share". . . , in light of the highly artificial levels and conditions of trade in wheat flour, the complexity of developments in the markets, including the interplay of a number of special factors, the relative importance of which it is impossible to assess, and, most importantly, the difficulties inherent in the concept of "more than equitable share"' (GATT, 1983).

Indeed, the concept of the 'equitable share' was not really meaningful, and certainly not well defined. It was never clear, and not at all agreed among all parties involved, whether the 'equitable share' is that part of the world market which an exporter would have in the absence of export subsidies,[43] or whether it is just the market share in a given reference period, even if that share had already been gained through export subsidies.[44] Exporting countries that felt they were competitive tended to believe in the former interpretation, while countries relying more heavily on subsidies subscribed to the latter view. Neither of the two interpretations is fully convincing. Determination of the market shares that would exist in the absence of export subsidies is impossible in any legally meaningful way. No two economists would agree on the counterfactual market situation that would exist in the absence of export subsidies. Reliance on a given reference period, independent of how market shares had been gained in that period, would in practice come close to freezing market shares. This would neither make economic sense, nor would it be in line with the philosophy of the GATT. All in all, the 'equitable share' was not 'a meaningful standard, mainly because the concept had no real intellectual anchor' (Hudec, 1993, p. 132). With these inherent difficulties it is no surprise that the attempt to amend Article XVI:3 of the GATT in the Tokyo Round through Article 10 of the Subsidies Code did not bear fruit. Market displacement was just as difficult to prove as violation of the 'equitable share'. The concept of price undercutting[45] turned out to be equally unworkable, not least because of the difficulty of finding reliable empirical information about prices actually paid in individual transactions. Moreover, economists would add, it is unlikely that price undercutting can ever be found because arbitrage will generally tend to make sure that the same product is not sold at different prices on the same market. Hence, the inability of the Code signatories to find a better and workable interpretation for Article XVI:3 demonstrated how weak this GATT provision was.

The problems the GATT 1947 was having with agricultural export subsidies were not limited to primary commodities. They also extended to processed products. The exception of Article XVI:3 was originally supposed to be strictly limited to raw materials. An interpretative note added to this GATT article explicitly defined the commodities falling under that exception as 'any product of farm, forest or fishery . . . in its natural form or which has undergone such processing as is customarily required to prepare it for marketing in substantial volume in international trade'. However, in practice this narrow definition was not really enforced. In the disputes over export subsidies for wheat flour, neither the complainants nor the panels even questioned whether flour was a primary product. This may reflect the fact that exporting countries, and certainly their milling industries, like to export flour instead of grain. However, it is surprising that the applicability of Article XVI:3 to flour was not even considered as it is certainly not necessary to mill grain in order 'to prepare it for marketing in substantial volume in international trade'.

It is even less necessary to process wheat into pasta before trading, which was why the US in one dispute complained about EC subsidies on the export of pasta products.[46] The case was rich in legal intricacies[47] but poor in outcome, since panel members could not agree. The majority found that pasta was an industrial product, and hence that subsidies on its export were prohibited. The dissenting panellist agreed with the EC contention that many countries had for a long time used export subsidies for the raw material content of their processed agricultural exports, and that this practice had implicitly become GATT-legal (though in legal terms the arguments were much more complex). In purely economic terms, much can be said for allowing a country to subsidize exports of processed products if, and as much as, exports of the incorporated raw materials are subsidized.[48] However, neither the text of the General Agreement nor the Subsidies Code allowed this to happen, although it had indeed become a universally adopted practice among subsidizing agricultural exporters.

Quite apart from the legal issues and economic inconsistencies, underlying the difficulties of settling disputes over agricultural export subsidies in the GATT was a fundamental disagreement among countries over whether it was appropriate for the GATT to impose effective constraints on the pursuit of domestic agricultural policies. In particular, if the EC, the defendant in most of the disputes over agricultural export subsidies, had ever accepted a ruling against it, this would have meant that domestic policies in the EC would have had to be changed as a result of a ruling from the GATT. The EC would have had to do away with one of the central elements in most of its agricultural market regimes, i.e. unlimited intervention buying at guaranteed prices. At the time, the EC was unwilling (and possibly unable) to do so, and other countries knew that.[49] The wheat flour and pasta cases were the occasions on which this became absolutely clear. As a consequence, GATT adjudication over agricultural export subsidies essentially broke down over these cases, and no serious attempt was later made to revive it.[50]

To sum up, the 'old' GATT failed to place any discipline on the use of export subsidies in agriculture. In contrast to the case of import measures, this was due to the agricultural exception of Article XVI:3.[51] This exception was thought to be 'realistic' in the sense of not outlawing a practice that countries were determined to continue to use. However, the way in which the exception was designed proved impractical because it could not be legally enforced. As a result, the battlefield shifted from legal procedures to world markets in the mid-1980s. Unable to settle their legal disagreements over Article XVI:3 and the Subsidies Code, the US and the EC engaged in a hefty subsidy war on world markets, much to the advantage of some big importers, but causing severe bitterness among the many smaller exporting countries.

The GATT and Domestic Subsidies in Agriculture

In the area of domestic subsidies, the GATT does not have any special rules for agriculture. This does not of course mean that domestic agricultural subsidies have not been contentious in the GATT. On the other hand, the number of GATT disputes over domestic subsidies in agriculture has been much smaller than in the areas of import barriers and export subsidies. In only three cases did countries complain about the effects of other countries' domestic subsidies in agriculture. One of these three cases would not occur until after the Committee on Trade in Agriculture had ceased to exist, during the negotiations of the Uruguay Round and with a major impact on their outcome in agriculture.

Domestic subsidies are not as such illegal in the GATT. Indeed, Article III:8(b) explicitly says that the requirement to treat domestic and imported products alike 'shall not prevent the payment of subsidies exclusively to domestic producers'. However, subsidizing countries have to observe certain precautions. If the subsidy results in increasing exports or reduced imports, it must be notified to the GATT, and if it leads to 'serious prejudice' to the interests of other countries, the subsidizing country 'shall, upon request, discuss . . . the possibility of limiting the subsidization' (Article XVI:1). Moreover, a country negatively affected by another country's subsidy can take the matter to the GATT if it feels that a benefit accruing to it under the Agreement (usually the value of an earlier tariff concession) is 'nullified or impaired'. As the subsidy concerned is not GATT-illegal as such, these complaints are called 'nonviolation nullification and impairment' cases (Article XXIII:1(b)). If successful, such complaints can result in a verdict requiring the subsidizing country to change its policy. In addition, importing countries have the possibility, under certain conditions, to levy a countervailing duty on the importation of products receiving subsidies in the exporting country. Countervailing duties have caused some friction in agricultural trade, in particular between the US and Canada. However, they will not be discussed here because they do not involve the legality of the domestic subsidy concerned.[52]

Two of the GATT disputes over domestic subsidies in agriculture dealt with production aids the EC was granting fruit processors to compensate them for the high prices they had to pay for their raw materials on the domestic EC market.[53] Only one of the two disputes resulted in a panel ruling.[54] In economic terms the underlying issue was essentially the same as in the pasta case: price support to farmers makes life unduly difficult for food processors unless the government finds some way to offset the disadvantage caused by high raw material prices. The panel implicitly accepted that view. However, the panel found that the subsidy to EC fruit processors had been more generous than necessary to compensate the disadvantage resulting from higher raw material prices. On that basis the panel ruled that the subsidy constituted nonviolation nullification and impairment. Although the panel did not explicitly say so, this ruling implied that a subsidy to fruit processors which just offset the price disadvantage would not have been a problem. The panel steered clear of the question of whether support to the growers of fruit violated GATT provisions, arguing only that it could not be said that this support to growers had worsened the competitive situation of imports by processors of canned fruit, the product complained about.

Interestingly enough, the third panel dealing with a domestic subsidy in agriculture would also avoid an explicit ruling on the question of whether the subsidy concerned had resulted in an unacceptable expansion of output of the farm product involved. The case was the oilseeds dispute between the USA and the EC, brought before the GATT in 1988. Because this dispute and the way it was successfully settled had a major impact on the agricultural negotiations of the Uruguay Round it will be discussed in more detail below in that context (see Chapter 7). Of course, this oil-seeds dispute could not affect thinking in the Committee on Trade in Agriculture because it came after the Committee had ceased to exist. However, at this point it is important to note that even in this important GATT dispute over a domestic subsidy, which in terms of its later effects on actual policy-making would turn out to be the most successful agricultural dispute in the history of the GATT, the central economic issue would not be settled, i.e. what is an acceptable level of a domestic subsidy to agricultural producers? The oil-seeds panel would find other (and as we shall see below not economically fully convincing) ways of arguing that the EC subsidy to oil-seed growers, paid through processors, constituted nonviolation nullification and impairment. But it would not provide guidelines which could be used to determine limits to the extent governments could pay domestic subsidies to their farmers. Given the magnitude and economic importance of domestic subsidies in many countries' agricultural policies, it was unfortunate that the GATT had not found a way to set constraints to their acceptability. In the absence of such constraints it could not be expected that there was much discipline in the way domestic subsidies were used in agriculture.

Outcome of the Committee's Work

Given the state of affairs in the application of GATT's rules to trade in agriculture described above, the Committee on Trade in Agriculture had a difficult task. There was no doubt that the GATT did not operate well in agricultural trade, though for different reasons in different areas. As far as market access was concerned, Article XI:2(c) was a potentially important exception for agriculture from the 'tariffs only' principle of the GATT. However, this exception was so narrowly circumscribed, and interpreted so strictly by panels in disputes, that very little use was made of it. On the other hand, this did not prevent countries from using other loopholes to protect their domestic markets effectively through a variety of non-tariff barriers to trade. In the area of export subsidies, Article XVI:3 afforded agriculture a notable exemption from the general prohibition of export subsidies. The constraints to exploiting this exemption had been weak to start with, and effectively they became even weaker when it turned out that they could not be strengthened in legal disputes. Governments were essentially free to subsidize exports as they saw fit for domestic policy purposes, and they used this freedom unashamedly. With regard to domestic subsidies, there were no special exemptions for agriculture, but then GATT 1947 rules were generally not very restrictive in this area, and governments used the wide scope for policy action generously. In the few disputes on domestic subsidies in agriculture, panels interpreted what there was in the form of rules rather strictly, and in one case this would have a significant impact on policy implementation in the country concerned. However, no general guidelines regarding acceptable levels of domestic subsidies in agriculture emerged from these disputes, and so there was no effective GATT discipline in this area either.

The Committee decided to organize its work in two parts. In part A, it was to examine all trade measures affecting market access and supplies. In part B, the CTA was to examine the operation of the GATT as regards subsidies affecting agriculture, especially export subsidies and other forms of export assistance. In both parts the examination was to be based on country notifications also called for by the Ministerial Declaration, and on papers prepared by the GATT Secretariat. The Committee was also to review relevant past cases of panels and working parties (BISD 30S/100–106).

The process of country notification produced a vast amount of paper in what was named the AG/FOR series of GATT documents, updating and to some extent completing the information collected earlier in the AG/DOC series.[55] All 41 participating countries and the European Community had to list, by product, all measures affecting exports and imports, and to state the GATT provisions on which they believed the measures were based. In particular, countries were also supposed to classify their measures according to their GATT status (including the categories 'provisions with special reference to agriculture in the General Agreement', 'waivers and derogations', 'lack of observance or application of certain provisions', 'particular interpretations of certain provisions', and 'not explicitly provided for in the General Agreement, e.g. variable levies, voluntary restraint agreements, long-term arrangements'). It came as no surprise that in this exercise countries were not strongly inclined to classify their measure as definitely or potentially GATT-illegal.[56] In the end this laborious process of notification under the CTA achieved not much more than some interesting but essentially inconsequential discussions in the Committee about individual countries' policies and their implications for agricultural trade.

More important were the discussions held in the CTA on how to deal with agricultural policies in the future, and its recommendations in this regard. Indeed, work in the CTA was in effect the start of the negotiating process in agriculture which later was formally initiated in the Uruguay Round. The text on agriculture in the Punta del Este Declaration launching the Uruguay Round in 1986 relied very much on wording developed by the CTA, and explicitly stated that the negotiating group responsible for agriculture would 'use the Recommendations [agreed by the CTA] . . . and take account of the approaches suggested in the work of the Committee on Trade in Agriculture . . .' (BISD 33S/24). The CTA remained active until shortly before the Uruguay Round proper and held its last meeting in April 1986. At that meeting, some delegates suggested that the CTA should become the negotiating group on agricultural matters in the Uruguay Round, and – though a decision to that effect was never formally taken – this is what effectively happened. As far as papers produced are concerned, work of the CTA culminated in two documents. One was the set of recommendations agreed by the Committee, and adopted by the CONTRACTING PARTIES at their 40th session, in 1984 (BISD 31S/10–12). The other one was a Draft Elaboration of these recommendations prepared, in consultation with the CTA chairman, by the GATT Secretariat.[57]

The recommendations of the CTA suggested that 'the conditions should be elaborated under which substantially all measures affecting trade in agriculture would be brought under more operationally effective GATT rules and disciplines', (including all the measures which in the past had effectively escaped GATT disciplines such as voluntary export restraint agreements (HVER), variable levies, unbound tariffs and export subsidies), and that 'approaches should be elaborated, as a basis for possible future negotiations' which might achieve this objective. The CTA also recommended that 'sanitary and phytosanitary regulations and other technical barriers to trade . . . [be] brought within the ambit of improved procedures aimed at minimizing the adverse effects that these measures can have on trade in agriculture'.

The Draft Elaboration of these recommendations also reflected alternative concrete approaches to reaching these objectives as they had been considered, on the basis of suggestions made by the countries participating in the CTA, in the discussions held in the Committee. No firm conclusions were drawn regarding the comparative merits of the alternative approaches. Instead, the rather different ways of dealing with market access and export competition which had been suggested by the participants were neutrally listed in the document. The text of the Draft Elaboration, therefore, reflected well the thinking on how to deal with agriculture as it stood on the eve of the Uruguay Round, and the large extent to which views on appropriate solutions still diverged among the parties concerned.

In the area of market access, the elaboration document reflected the amount of time and attention spent in the CTA on ways to strengthen disciplines on quantitative restrictions, variable levies, minimum import prices, state trading, and other NTM. Much thinking by participants went into identifying and discussing various options for improving the functioning of the GATT Article XI:2(c)(i), and a clear preference was discernible to bring not only outright quantitative restrictions but also other NTM under disciplines along the lines of that GATT provision. Domestic supply controls and minimum access commitments as prerequisites for the application of NTM appeared, to many participants, as an attractive way of disciplining barriers to market access. Rather different approaches to making Article XI:2(c) more operational were suggested. On the one hand, a much stricter definition of the 'restriction of production' criterion was proposed by some participants. On the other hand it was suggested that 'disciplines should be realistic and should take full account of the specificity of the agriculture sector . . . so that restrictions outside Article XI could be brought within effective GATT rules and disciplines'. However, at the opposite extreme the proposal was also made by other participants in the CTA to adopt a 'tariffs only' approach which would entail a phasing out of all NTM, a deletion of all but paragraph 1 of Article XI, and a binding of all previously unbound tariffs. This approach, later actually adopted as 'tariffication' in the Uruguay Round, was at the time considered to be rather unrealistic by most participants in the CTA.

As far as export competition was concerned, the CTA recommendations also reflected the wide divergence of views, and suggested two rather different broad approaches, i.e. either improvements in the existing GATT framework (Article XVI) or a general prohibition of export subsidies, subject to carefully defined exceptions. Proponents of the improvement school in the CTA suggested various ways of defining the infamous 'equitable share' more operationally, for example by calculating, or agreeing in negotiations, predetermined market shares, or by establishing new criteria such as the share 'which might reasonably be expected to prevail in the absence of the subsidy' or which 'is inconsistent with the objective of satisfying world market requirements of the commodity concerned in the most effective and economic manner' or 'is acquired . . . with the effect of depressing prices'. This latter school also came up with various suggestions regarding the definition of the 'special factors' mentioned in Article XVI:3 which had caused so much headache to past panels, and with new ways of determining serious prejudice. The prohibition school in itself was widely divided. On one extreme, the suggestion was made by some participants that 'all existing export subsidies should . . . be phased out and new export subsidies should be prohibited'. The related wording that 'existing export subsidies would be permitted temporarily so long as a schedule and phase-out rules are established for each subsidy and the export quantity is gradually reduced' sounds like a precursor to the approach finally adopted in the Uruguay Round. At the other end of the spectrum, some proponents of the prohibition school also had rather generous proposals on exceptions, such as exempting all subsidies from the prohibition which were accompanied by a given increase in market access in the exporting country concerned, or which were consistent with a given negotiated maximum self-sufficiency ratio of the exporting country, or which were producer-financed.

On sanitary and phytosanitary measures the CTA did not make much headway. One suggestion was that countries should have 'to justify, or accept some measure of review of, the grounds on which . . . imported products are subject to more stringent rules or requirements than those applied to the domestic products'. It was also said that 'any improvement in the existing procedures would appear to depend to some extent on whether the opinion of suitably qualified or experienced trade and technical experts could be brought to bear on such questions in a GATT dispute settlement or consultation context'. Another possibility considered was to require countries to compensate with concessions on other products if they had nullified concessions by erecting technical barriers to trade. Compared with the later Uruguay Round agreement on sanitary and phytosanitary measures, the

CTA work in this area now appears rather conservative and narrowly confined.

The CTA also recommended a process of regular review and examination of agricultural trade measures and policies, to be based on a continued system of notification along the lines of the AG/FOR series of documents established and reviewed by the Committee. The usefulness of such a review process, it was said, would very much depend on the nature of commitments which might eventually be undertaken by governments. Given the widely diverging views in the Committee on whether any firm commitments should be agreed, and what the character of such commitments should be, the final remark on the review process was that it was 'not possible at this juncture to comment otherwise than in rather general terms on the scope and modalities of a multilateral review and examination of national policies and measures in the GATT'.

In his 'non-paper', the CTA chairman (Aart de Zeeuw) referred to the difficulties which the Committee had faced by stating that 'the problems we have to deal with are deep-seated and are not going to be easily solved, especially because they are highly political. The need to find lasting solutions are [*sic*] as pressing now as they were [*sic*] when the Committee embarked on its work. . . . Since domestic policies as such are not negotiable the approach proposed and accepted in the GATT Committee would involve rights and obligations in GATT being a function of the manner in which internal policy is conducted'. The non-paper did not propose concrete solutions, but what it contains in the form of more general suggestions illustrates, viewed from the perspective of ten years after the CTA, how far the Uruguay Round Agreement on Agriculture went beyond past approaches on agriculture in the GATT.

In retrospect it is difficult to say precisely how useful the CTA was. Its work did not result in any concrete action which would have improved the state of affairs in agricultural trade at the time. In that sense the CTA did not achieve any direct result. On the other hand, the largely inconclusive debates in the Committee have served the important purpose of making it clear, one more time, how important it was to make progress, and how difficult, if not completely useless, it would have been to try to fiddle with marginal improvements to the then existing GATT rules for agriculture, as embodied in Articles XI, XVI and XX. In that sense the CTA made an important contribution to preparing the intellectual ground for the Uruguay Round negotiations on agriculture. It mainly considered a 'rules approach', which would have amounted to clarifying and tightening GATT rules on agriculture, by rewriting and augmenting GATT articles. How little such an approach might have achieved became clear during discussions in the CTA. At the same time, some of the commitments agreed on agriculture in the Uruguay Round had already been suggested by some parties in the CTA. In particular, the 'tariffs only' approach finally prevailed in the Uruguay Round, minimum access commitments made their way into the Agreement on Agriculture, and a phasing down (though not a phasing out) of export subsidies was eventually agreed. However, when such suggestions were made in the CTA, they were considered largely unrealistic and far from something which could possibly be agreed among all participants. Moreover, domestic support commitments as also later bound under the Agreement on Agriculture did not figure at all in the work of the CTA.

For the observer with an interest in history, and for the critics of the Uruguay Round Agreement on Agriculture, it is revealing to compare the deliberations of the CTA with the outcome of the Uruguay Round in agriculture. Seen from the perspective of the period immediately before the launching of the new GATT round, and judged against the background of the difficulties experienced in the CTA to even narrow down the wide divergencies of views on which approach to dealing with agricultural trade should be adopted, the Uruguay Round Agreement on Agriculture must appear as a major step forward. What a difference ten years of growing difficulties with traditional agricultural policies, worsening market conditions in international trade, and serious negotiating efforts can make!

7 The Uruguay Round Negotiations[1]

1 BACKGROUND TO THE URUGUAY ROUND

It must take stamina to be a trade negotiator. The Uruguay Round of trade negotiations was launched in September 1986 with the adoption of the ambitious agenda of the Punta del Este Declaration. It ended seven and a half years later with the signing of the Final Act in Marrakesh in April 1994. Not all of that time was taken up in active negotiations: as with the Tokyo Round, many months were spent waiting for a resolution to a few particularly sticky issues. The original intent had been to complete the negotiations by December 1990, but the four-year timetable began to look wildly optimistic as successive deadlines came and went.[2] Towards the end, the very act of bringing the Round to a conclusion began to be important to preserve the credibility of the multilateral process. When the Final Act had been approved the relief among those concerned with the future of the multilateral trade system was palpable.

Agricultural issues above all others were responsible for much of the delay. Intense discussion of those issues involved only a few of the negotiating parties. The inability of the EC and the US to reach agreement on the treatment of agricultural trade was the most obvious hurdle to be overcome. It was the proximate cause for the delay in reaching a 'mid-term' agreement in December 1988 and again in reaching a conclusion at the 'final' ministerial session in December 1990. Lingering disputes over agriculture even threatened the final push for agreement in November 1993. But unlike the previous two rounds, other countries with a strong interest in the outcome of the agricultural talks were not content to sit on the sidelines. In particular, a group of 14 agricultural exporters, the Cairns Group, participated actively from the start and made sure that the US and the EC did not again sweep agriculture under the carpet.[3]

In the event, the Uruguay Round represented a milestone for the multilateral trading system. An agreement of unprecedented scope emerged to guide the trading world into the next millennium. A fail-

133

ure to have finished the Round would have had a traumatic impact on the multilateral system as well as serious consequences for international trade. Costly trade wars, poisoned political relationships, and more difficult cooperation in other areas of international economic policy could have followed. Perhaps the rules that existed could have been enforced more consistently with an improved dispute settlement mechanism. Perhaps regional and bilateral trade arrangements would have continued the progress towards lower trade barriers. Perhaps these benefits could have been extended to other regions in a network of free-trade areas. Perhaps tensions between these trade blocs could have been held at bay to minimize the risk of trade wars. But it seems unlikely that any of these alternatives would have made up for the bitterness and disillusion that would have followed a failure to reach an agreement in the Round. Instead, as at the end of the Kennedy Round, the multilateral process emerged strengthened and invigorated, with a more permanent institutional base and a new set of rules governing an expanded portion of world commerce.

The Origins of the Round

The Uruguay Round had its origins in the early 1980s, when the US Administration began to canvass support for a comprehensive set of negotiations that would restore purpose and integrity to the multinational trading system.[4] The economic recession following the second oil crisis of 1979–80 had left developed countries desperate for remedies to the twin problems of slow growth and chronic unemployment. The debt burden constrained growth in the developing world like a massive economic hangover from the previous decade. Only in parts of Asia was the economic picture bright. But the growing success of the North-East Asian economies in pursuing export markets in Europe and North America only added to the pressures for protection in the West. The US Congress reflected these concerns and began to seek unilateral solutions, including trade actions tied to bilateral trade balances, which threatened the foundations of the multilateral trade system.

The response of the Reagan Administration was to call for a GATT ministerial meeting in 1982, a rare occurrence outside the framework of specific negotiating rounds. The timing was not propitious, and the meeting exacerbated rather than reduced the tensions in the trade system. It ended in fractious disarray, with no clear support for the US position and no agreement as to what should be the next step. The GATT system seemed incapable of reacting to the changing trade environment. To be sure, the export success of Japan had increased the interest of that country in a strengthening of multilateral trade rules. But the EC showed little enthusiasm for another GATT round so soon after the Tokyo Round, and did not relish the inevitable attack on the CAP which would characterize further talks on agriculture.

The US, however, remained adamant that new talks were needed. The inconclusive 1982 Ministerial Meeting had shown the difficulty of getting agreement on improvements in the trade system. But the emergence of a new set of problems gave the necessary push to trade officials and set a stamp on the international agenda.[5] These new issues included trade in services, in which the US was the leading exporter, the protection of intellectual property rights such as copyrights and patents, and investment policies of overseas countries which often put constraints on the activity of US firms abroad.[6] Coupled with continuing concern about the restrictions on the sale of farm goods abroad, and a general sense that the US was losing ground to other countries in many manufacturing sectors, these 'new' trade concerns provided an expanded agenda for the prospective GATT Round.

As if to show that other options existed, the US also began at that time to explore the notion of regional trade agreements and free-trade pacts with other countries. In February 1982 President Reagan announced the Caribbean Basin Initiative. The determination of the US throughout the postwar period to avoid regional preferences, itself a reason for the nondiscrimination cornerstone to the GATT, was weakening under the political need to keep the Communist influence in the Caribbean region at bay. The US found itself emulating the preferential trade regimes that had characterized the EC's relationships with the Mediterranean littoral. Congress was less than enthusiastic about the trade provisions of the new policy, but eventually passed the necessary legislation embodied in the Caribbean Basin Economic Recovery Act in July 1983, after the invasion of Grenada had focused attention on the strategic importance of the region.

The twin-track policy of the Administration, pursuing negotiations on both bilateral and multilateral fronts, continued during the run-up to the Uruguay Round. Discussions began on establishing a free-trade area with Israel following a state visit by Prime Minister Shamir to Washington in 1983. These talks eventually led to the signing of the US–Israel Free Trade Area Agreement in April 1985.[7] In March 1985 President Reagan met with Prime Minister Mulroney at the 'Shamrock Summit' and signed a Declaration on Trade in Goods and Services designed to set the scene for the negotiation of a US–Canada Free

Trade Agreement.[8] The world's largest bilateral trade flow, between the US and Canada, was thus to benefit from regional preferences.

Setting the Agenda

The agenda for the new GATT round reflected a mixture of the old and the new in trade concerns. The round was designed in part to bring into greater conformity with the GATT two of the sectors which had been out of the mainstream, namely textiles and agriculture. Textiles trade had strayed perhaps furthest from the principles of nondiscrimination and tariffs-only protection. World trade had become covered by a network of quantitative trade restrictions under the Multifibre Arrangement.[9] The removal of the MFA was high on the list of priorities for developing countries. Similarly, from the start of the Uruguay Round it was clear that there had to be a focus on agriculture, in an attempt to correct the problems identified in the previous chapter of poorly defined and ineffective rules. To these two elements of 'old business' were added three new negotiating areas: trade in services, which had been growing fast but which lay entirely outside the rules of the GATT; Trade Related Intellectual Property Rights (TRIPS), in particular the perceived need by industrial countries to see that developing countries implemented patent and copyright protection; and Trade Related Investment Measures (TRIMs), such as export performance and domestic purchase conditions attached to foreign direct investment activities. In addition, there was general agreement on the need to strengthen the subsidy and anti-dumping provisions, improve the dispute settlement process, and improve the coordination between trade matters and those relating to monetary and investment policy at the international level.

Getting agreement on this agenda among all the GATT contracting parties was not easy. Developing countries, led by Brazil and India, objected to the inclusion of services in the international trade agenda, fearing that US, European and Japanese financial and other service institutions would be able to dominate their domestic markets if allowed liberal access. The same caution was expressed by countries concerned that the adoption of stricter rules for the protection of intellectual property would give established US and other firms the edge in trade in software, films, pharmaceuticals, and other copyright- and patent-intensive activities. Rules limiting the investment policies of developing countries were also seen as inimical to the interests of those countries, and an intrusion on their autonomy. The liberalization of the textile

quotas and the regulation of excessive protection in agricultural markets was in part a *quid pro quo* for developing countries, which would
presumably benefit from expanded exports. Japan was a keen supporter
of the broadened agenda, in particular those elements of it, such as
strengthening the dispute settlement mechanism and clarifying the rules
for contingent protection, that would guard the multilateral system from
the impact of US Congressional action.[10] The EC was not unwilling to
negotiate on the full agenda, many items of which were also on the
internal agenda for the completion of the single market.[11]

The Timeline of the Negotiations

Before the Uruguay Round negotiations could get under way, countries had to agree on the broad objectives of the negotiations, exchange
ideas as to the approach to be taken and decide the way in which
the negotiations themselves should be organized. The definition of the
objectives was the subject of tense negotiations in the lead-up to the
Punta del Este meeting. The relation between the goods negotiations
and those in the area of services caused the most problem. Brazil and
India had argued for complete separation of the two areas, concerned
that they might be forced to yield in the service area to get any advance in other parts of trade. An informal meeting of trade ministers
in Stockholm, in June 1985, reached a tentative compromise to negotiate on goods and on services in 'parallel but separate' talks. In September, countries set up a 'Preparatory Committee', with GATT
Director-General Arthur Dunkel as the Chairman, and proceeded to
discuss the objectives of the Round. The work of this Committee dragged
on into 1986, until Dunkel forced the issue by setting the date of September 1986 for the ministerial meeting to launch the Round. A group
of 40 countries, concerned about the lack of progress, met in July in
Geneva for an intensive effort to define the objectives.[12] The outcome
of this meeting was an agreement not to link concessions in goods
with those in services, and an understanding on agriculture that removed the reference in an earlier draft to removing export subsidies.[13]
The group of informal negotiators, now numbering 47, took the draft
declaration to the final Preparatory Committee meeting in July, though
the EC found it necessary to abstain from a formal endorsement.[14]

   The formal negotiations began in September 1986 with the launch
of the Round at Punta del Este, Uruguay. The deliberations were to be
conducted on two parallel tracks, the one to negotiate on matters relating to trade in goods and the other to tackle the new issue of services.

The goods negotiations group (GNG) in turn set up a series of 14 committees to conduct the talks on the various topics. The GNG, together with the group charged with negotiation on services (GNS), reported to the Trade Negotiations Committee (TNC). Each committee could either meet at the level of officials, led by the respective ambassadors to the GATT, or on occasions at the ministerial level. The agricultural negotiations were to be handled by a Negotiating Group on Agriculture (GNG5), which was to be chaired by Aart de Zeeuw, the former Director-General in the Dutch Ministry of Agriculture who had also been Chairman of the 1982 GATT Committee on Trade in Agriculture.

The first step in the negotiations was to lay out a framework for the talks. Each negotiating group discussed the possible modalities for negotiation. The agricultural talks at once became controversial when in July 1987 the US unveiled a dramatic proposal for eliminating all trade-distorting farm programmes over a ten-year period. For agriculture, this first phase produced more heat than light. Other negotiating committees made somewhat more progress. In early 1987, Clayton Yeutter, the US Trade Representative at the time, floated the idea of an 'early harvest' of results from the Round to be prepared for the 'mid-term review' which had been called for Montreal in December 1988. This review would help to define the timetable and agenda for the remaining two years. This effort did not go according to plan: the initial stage in the Round culminated in the suspension of the negotiations at the Montreal meeting, with the Cairns Group insisting on progress in agriculture before it would let the talks proceed.

The negotiations were rescued in April 1989, when countries finally agreed to a compromise wording on the objective of the agricultural talks and completed the 'mid-term' package of trade measures. Thus began a second phase of the negotiations, which consisted of an elaboration of the negotiating ideas by each (major) participant, with the intention of leading to a common document on which all parties could focus attention. Despite attempts to get consensus, the 'final' negotiations in Brussels in December 1990 collapsed, largely as a result of the impasse on agriculture. Once again, the Cairns Group indicated its unwillingness to settle for a weak compromise on agriculture. Without an agricultural component, all other parts of the negotiations were put on hold.

Agreement on the structure of an agricultural package did not occur until February 1991, after the EC had decided upon substantial modifications in its own internal agricultural policy, and was therefore able to live with the changes implied by a GATT agreement on agriculture. This ushered in the third and final phase of the GATT negotiations.

The drafts of all the individual agreements, including an attempt at a new agricultural package, were incorporated into the 'Draft Final Act' of December 1991, usually called the Dunkel Draft. Countries generally accepted the Dunkel Draft as a basis for a final agreement, and proceeded to negotiate on the details and on the accompanying package for market access improvements. The agricultural component was modified in important respects in November 1992 by the Blair House Accord between the US and the EC, and later refined by tense last-minute negotiations in Geneva in December 1993. Finally, agreement was reached on December 15, as delegates from 117 countries heard the Director-General of the GATT, Peter Sutherland, declare 'I gavel the Uruguay Round concluded'.[15] The time between the end of the negotiations in Geneva and the signing of the Final Act in April 1994 in Marrakesh was devoted to a clarification and verification of the individual country schedules that comprised the details of the implementation of the Agreement.

Subsequent to the signing of the Uruguay Round Agreements, most countries had to ratify them under their various domestic legislative procedures. Even this step provided some last-minute drama. The US Congress, never enthusiastic about multilateral trade rules and stronger institutions, extracted some promises from the administration before finally giving President Clinton the ratification he sought in December 1994.[16] The newly elected lower house of the Japanese parliament quickly followed suit. The EC countries argued internally about whether the expanded scope of the Round to include services and intellectual property had not taken it beyond the jurisdiction of the Community and hence required national ratification as well as that by the Council of Ministers. However, a timely pronouncement from the European Court of Justice in November 1994 declared 'shared responsibility' for the expanded menu of trade issues, and allowed the ratification process to proceed. By Christmas 1994 the twelve member States had ratified the Agreement, and the Council was able to commit the EC to the Uruguay Round Agreements and membership of the World Trade Organization. The Uruguay Round Agreements went into effect, and the WTO was born, on 1 January 1995.[17]

2 THE AGRICULTURAL NEGOTIATIONS

The agricultural negotiations largely determined the pace and progress of the Round. The first phase (September 1986 to December 1988) consumed over one half of the four years originally set aside for the Round in discussions over what should be negotiated, but in the end yielded no conclusion as to how to proceed. The second phase (April 1989 to December 1990) was taken up with the efforts of the negotiators to come up with an agricultural agreement which would allow victory to be declared, but once again ended without success. The third stage (February 1991 to April 1994) represented the main negotiation phase, with all countries finally on the same page. The story of the negotiations is that of the concepts and ideas which were discussed from the start of the Round and which found their way into the agreement, and of the political roller coaster of alternate stalemate and breakthrough which gave an element of theatre to the agricultural talks throughout the various stages.

Phase One: From Punta del Este to Montreal

The Punta del Este Declaration of September 1986 established the objectives of the negotiations and defined their structure. Agricultural negotiations were to be an integral part of the negotiation on goods trade, but conducted in an agricultural committee. The sector was singled out for specific mention in the Declaration. GATT contracting parties agreed on the 'urgent need to bring more discipline and predictability to world agricultural trade' and took as the aim of the negotiations 'to achieve greater liberalization of trade in agriculture and bring all measures affecting import access and export competition under strengthened and more operationally effective GATT rules and disciplines' (BISD33S/24). The significance of these objectives for agriculture cannot be overemphasized. They broadened the scope of negotiations beyond the traditional areas of trade measures and border instruments. For the first time, domestic agricultural policies were to be the subject of international negotiations, though at that time it was unclear as to whether their level and instrumentality as well as their trade effects were on the table. The Declaration gave the necessary political commitment for negotiators to begin their deliberations but did not indicate the nature and form of the negotiations themselves.[18] At a meeting in January 1987 the negotiators called for countries to table suggestions as to how to conduct the agricultural negotiations (BISD 33S/40). The series of papers tabled in the second half of 1987 was the response to this request.

The 1987 US Proposal

The US paper arrived first and set an ambitious agenda for the talks. The paper advocated a sweeping approach to achieving the Punta del Este aims – eliminate over ten years all the trade-distorting policies plaguing agricultural markets. Participants were invited to agree to 'a complete phase-out over 10 years of all agricultural subsidies which directly or indirectly affect trade', to 'freeze and phase-out over 10 years the quantities exported with the aid of export subsidies' and to 'phase-out import barriers over ten years' (GATT, 1987a). The primacy of dealing with agricultural subsidies was a significant innovation. Whilst paying lip service to the notion of reducing trade barriers and curbing export subsidies, the main thrust of the US proposal was clearly on the reduction in levels of support and protection. The first 'tier' of an agreement would include the measurement of levels of support and the negotiation of a schedule for their elimination. The monitoring of the level of support was to be done through a quantitative indicator, such as the Producer Subsidy Equivalent (PSE) that had been popularized by the OECD. The second tier would involve mutual agreement on the 'specific policy changes' which would be submitted by each country 'to meet its overall commitment of scheduled support reductions'. Countries would present a 'country plan', for other contracting parties to scrutinize and approve. The process of domestic policy formulation would not only be made transparent but be subject to international discussion, though it was not entirely clear how much direct influence would be exerted at the stage of approval of country plans. GATT rule changes were not discussed in the proposal: instead there was a suggestion that 'negotiation should begin on the changes necessary to GATT rules to reflect the trading environment that will exist at the end of the transition period' i.e. in the year 2000. These rule changes would of course be less contentious if agricultural support were to be eliminated and domestic policy instruments were changed to only those that had no (or minimal) trade impact.

The US proposal did three things for the progress of the negotiations. First, it raised the sights of the negotiators by suggesting a radical departure from piecemeal tinkering in favour of the establishment of a lofty goal and a definite time path to achieve the objective. In this respect it captured the negotiating high ground and served notice on other countries that the US was prepared to 'think big'. A more circumspect position might have been to suggest the removal of, say, one half of the trade-distorting protection, itself a daunting task.[19] But the

Reagan Administration had shown its willingness to take a similar broad-brush approach to domestic farm legislation. A draft of the 1985 Farm Bill had already suggested eliminating domestic farm policies in the US.[20] Although this version did not prevail, the approach showed a calculated disregard for the principle of marginalism in the improvement of policies. In this respect the 'zero-2000' proposal got the agricultural talks off to an ambitious if controversial start.

The second contribution of the US paper was to throw support behind a new technique for negotiation on agriculture, and indeed in any sector of the economy. Traditional negotiations in the GATT had taken three forms: changes in trade rules that apply to all countries, and taken without explicit regard to the impact of those rules on individual countries; requests and offers for particular trade 'concessions', usually in the form of increased market access, negotiated bilaterally and then made available to all through the most-favoured nation doctrine; and 'across the board' cuts in tariffs, either using simple percentage reductions or formulae that weight the cut by the height of the existing tariff.[21] The novelty introduced in the US proposal was to use a proxy for the level of support and then to negotiate on this proxy. Countries could choose their own path to salvation, subject to approval – it was the end product that mattered rather than the modalities. This technique, based on the use of an aggregate measurement of support (AMS), received considerable attention in the first and second phases of the negotiations, though its significance was much reduced by the final stage of the talks. Its manifestation in the Final Act is but a shadow of its role in the earlier parts of the Round.[22]

The third contribution of the US proposal was to ensure a head-on conflict between the US and the EC. There was no possibility of the EC agreeing to the removal of all trade-distorting support for agriculture, however beneficial this might have been for world trade flows and for international resource allocation. The idea that the EC could ever allow free access for agricultural imports into its internal market was too much to contemplate. The existence of trade barriers to third-country trade was seen as essential to the very definition of the EC, and the maintenance of agricultural trade barriers was held to be essential to operation of the CAP, which relied much more on trade instruments for domestic price and farm income support than did the US or other major trade partners. The delicate political compromises that supported the CAP would be damaged by such a recasting of the method of income support. In particular the budget cost of a move to direct payments would have been insupportable as a Community expense, and would probably have implied a 're-nationalization' of the CAP. The US may have held the high ground, and been able to defend its position with logic and principle: the EC had to consider political realities and the limitations imposed by its own internal arrangements.[23] The US proposal was hardly a recipe for a rapid solution to the talks, and indeed this confrontation between high principle and low politics kept the talks essentially stalled until after the Montreal meeting in December 1988.

The 1987 EC Proposal

The European Community countered the same month with a proposal of its own. It was both more traditional in approach, emphasizing the 'root problem effecting world agricultural markets, i.e. the imbalance between supply and demand' rather than the distortions created by domestic policies, and more limited in scope in seeking a 'reversal of the present trend towards structural disequilibria' (GATT, 1987b). The EC's two-stage plan proposed to deal with short-term issues of world prices (which were at that time depressed) before moving to a period of support reduction and strengthened rules. The first stage would attempt 'to ease the situation on worst-affected markets' and to restore 'healthy' market conditions by means of 'a concerted reduction of support'. Two parallel activities would be undertaken during this first stage: 'emergency actions' in the market for cereals (and cereals substitutes), sugar, and dairy products, and undertakings to reduce support to avoid the 'exacerbation of existing imbalances'. However, no policy changes would be implied during this first stage. In a second phase, GATT members would undertake to carry out 'a significant, concerted reduction in support coupled with a readjustment of their external protection'. This reduction 'could be backed up by aid to farmers designed to offset the loss of earnings occasioned by the new arrangements'. The EC envisaged that bindings of 'maximum levels of support, protection and export compensation' would be possible at the end of this second phase. Implementation of these plans would 'allow the GATT rules and disciplines applicable to agriculture to become genuinely operational'.

The EC paper was initially welcomed in Washington for the extent to which it broke ground with previous EC positions, though US negotiators in Geneva quickly turned to condemnation of its shortcomings.[24] It was indeed a notable document, breaking new ground (for the EC) though still betraying the internal conflicts on the subject within the

Commission and the member states. The fact that the EC was prepared to accept any international disciplines on the level of support afforded by the CAP was itself a major step. The inability of the EC to contemplate a full reinstrumentation of its policies was understandable, though the CAP was certainly in need of different instruments.[25] The implied call for market management to firm up world prices in the short run provided a link with the past, when the EC had often argued for the reliance on commodity agreements to shore up world markets – and incidentally make the CAP easier to run.

The 1987 Cairns Group Proposal

It was to counter the tendency of agricultural trade talks to degenerate into a bilateral US–EC confrontation that the Cairns Group had been formed. The 14 countries, which met in the Queensland town of Cairns in August 1986, identified themselves as 'non-subsidizing' traders in agricultural products that had been particularly badly hit by the export subsidies of the US and EC. They shared a disillusionment with the way previous GATT rounds had come to a marginal deal on agriculture for the sake of reaching closure, but which had done little to correct the underlying weaknesses of the agricultural markets into which they sold their produce. The members of the Cairns Group decided to join forces to present a united voice as a counterweight to those of the EC and the US. It was also assumed at the time that this third voice would be a moderate and compromising element, able to be supported by other countries, and perhaps even by the two main protagonists.

The Cairns Group's strategy was to come forward with its own proposal, rather than just react to those of the 'agricultural superpowers'. This it did in October 1987. The Cairns Group was slightly less radical than the US in its approach to the agricultural negotiations, but it also delivered a frontal attack on agricultural protectionism. The Cairns Group proposal suggested a three-stage plan which would effectively allow for a freeze on protection, followed by a scaling back of support, until finally a new set of rules could be instituted which would in effect ban the use of trade-distorting policies (GATT, 1987c). The first stage of the Cairns Group plan would have concentrated on 'early relief' measures which would have prevented any increase in protection in agricultural markets, frozen existing export and production subsidy levels, and taken a 'concerted multilateral first step' to improve agricultural trade by means of an 'early-harvest' package of subsidy cuts and access increases. The medium-term reform programme was similar in concept to that of the US proposal. Countries would undertake commitments 'to reduce and eliminate trade distorting policies in the form of country schedules', meet 'targets of reduced levels of overall support' and target 'those support and policy measures which contribute most heavily to trade distortions' for priority attention. Monitoring of this process was to be facilitated by the 'establishment of a measure of aggregate support capable of capturing the diverse policies' that countries employed, and the Group suggested the use of a PSE-type measure. The long-run framework of new rules to which the agreement would lead would include the prohibition of all import barriers 'not explicitly provided for in the GATT', the elimination of all waivers and derogations and the 'binding of all tariffs on agricultural products at low levels or zero'. There should in addition be a prohibition on the 'use of all subsidies and other government support measures . . . having an effect on agricultural trade'. The new rules would thus have brought agriculture fully into the GATT with a ban on non-tariff barriers and export subsidies, and a required reinstrumentation of national agricultural policies with a set of domestic programmes less injurious to trade.

The initial paper from the Cairns Group was a remarkable mixture of a clear objective, flexible means to attain that aim, and a short-term commitment to start off the process. While it did not receive as much attention as the US and (later) the EC proposals, it established the Cairns Group as a serious player in the game.[26]

Other Proposals

Importing countries also added their ideas as to the conduct of the negotiations. A paper by the Japanese negotiators saw little merit in the notion of negotiating support reduction, and 'no need for a comprehensive aggregate measurement of the level of protection and support' (GATT, 1987d). The main blame for the problems of world agricultural markets lay with the exporters. 'It should be recognized that increased subsidized export of agricultural products . . . has seriously distorted the world trade in agriculture'. The present restrictions on export subsidies were declared to be 'not practicable', and the proposal argued that the basic rule prohibiting the use of export subsidies . . . should equally be applied to the export of agricultural . . . products'. The paper went on to argue that 'domestic subsidies other than export subsidies' which were being used 'as a part of agricultural policies' should 'clearly be distinguished from export subsidies'. The paper did not however claim that domestic policies are outside the purview of trade talks. It merely suggested that they be judged on their internal, not their external, appropriateness.

The Nordic countries also put in a proposal for the negotiating framework (GATT, 1987e). This paper was remarkable in its willingness to contemplate support reductions, in spite of the high levels of support that prevailed in these countries and their inability to include agricultural trade in their deepening economic ties with the EC.[27] A discussion paper circulated by a group of food-importing developing countries was never formally presented as a negotiating proposal, but it served to remind delegates of interests other than those of the major and middle-rank agricultural powers.[28]

The year following the tabling of the initial proposals was spent more in posturing than in negotiation. The Negotiating Group on Agriculture met seven times during the year, and considered the proposals at length. Working Parties were set up to consider the issue of the aggregate measurement of support and improvements to the sanitary and phytosanitary rules. Each position was elaborated, and additional papers were tabled which explained these national positions to the other parties. In April 1988 the Cairns Group presented a paper which reflected the outcome of their meeting at Bariloche, calling for the immediate implementation of a two-year programme of support reduction as a 'down-payment' while the negotiations proceeded (GATT, 1988b). Later, this proposal was further defined as a 10 per cent reduction in support for the years 1989 and 1990 (GATT, 1988c). The force of the proposal was, however, weakened by the difficulties that Canada had with the potential impact on its own politically sensitive dairy industry. The EC expanded on its proposal to take short-term measures by freezing levels of support at their 1984 levels, thus ensuring full credit for CAP reforms since that date and taking as a reference period the time when the dollar was unusually strong. In October 1988 the Japanese negotiators gave cautious approval to the use of an AMS for monitoring protection, and the US agreed to contemplate a short-term freeze if others would agree to the longer-term objective of an elimination of support (GATT, 1988d).

As the scheduled mid-term review approached, views on the conduct of negotiations, however, tended to polarize rather than coalesce. The US continued to insist on an advance commitment to the elimination of trade-distorting subsidies, and the EC steadfastly declined to be drawn into such an obligation, declaring it to be far beyond the agreement at Punta del Este. The EC labelled the US 'zero-option' proposal unrealistic, but did not offer its own view of a longer-run objective. With the US election imminent in November 1988, the EC appeared to have little interest in negotiating with the outgoing Administration. The US in turn showed no willingness to compromise its principles, even though it was thereby holding up the talks. Domestic policy moves, such as the relaxation of acreage set-asides in 1988 (following a short corn crop) were taken to put pressure on the EC, as was the expansion of the Export Enhancement Program (EEP) which attempted to dispose of surplus products on markets where the EC (through its own export subsidies) had secured a growing market share.

The Montreal Meeting

Matters reached a head at the time of the mid-term ministerial meeting, held in December 1988 in Montreal. The US 'early harvest' called for progress in the areas of dispute settlement, the functioning of the GATT system (FOGS), services, tariff reductions, and agriculture. Developing countries looked for progress on reducing barriers to tropical products. The EC pushed the notion of 'globality', that is the need to push ahead on all fronts simultaneously, presumably to avoid the pressure for an agricultural breakthrough. An initial agreement on services materialized, and constructive progress was made in the other areas, such as the need for a regular review of the trade policies of member countries (the Trade Policy Review Mechanism). Access for tropical products to developed country markets was improved, largely through the expansion of the Generalized System of Preferences. The main issue outstanding was agriculture, where intensive negotiations in Montreal failed to find a solution to the unbridgeable gap between the United States' insistence that a commitment be made to 'eliminate' support and protection for agriculture and the EC's willingness to concede only that it should be 'reduced'.

The Montreal meeting of the TNC was meant to be the time to bring together the individual negotiating groups to identify the progress made and to define the agenda and the timetable for the final two years. Difficulties in some areas would be resolved as a part of a package with other areas. Unfortunately the reverse happened. The deadlock in agriculture spilled over to the other sectors, even where progress had been good. Without an agricultural component, agreement on the mid-term 'package' was not possible. In the end it was the Cairns Group delegates, led by Argentina, who decided to call a halt to the TNC meeting. US and EC negotiators could not find that elusive form of words that would at the same time signal to US agricultural interests that the EC was prepared to abandon farm price support in the long run and to EC farmers that the Commission had fended off the frontal attack by American negotiators.[29] Talks were formally suspended in the wake of the Montreal meeting, and informal discussions were held between the GATT Director-General and the negotiating parties to come up with a formula which would allow the Round to continue.

Phase Two: From the Mid-Term Review to Brussels

The deadlock in the negotiations was finally resolved in April 1989, with the adoption of the mid-term progress report (GATT, 1989a). The mood changed from the pessimism of the Montreal walkout to the bright optimism of a fresh start. The negotiating teams themselves had changed. In the US, the incoming Bush Administration appointed Carla Hills to be US Trade Representative, and Clayton Yeutter moved to be Agriculture Secretary. Dan Amstutz was not reappointed as the lead negotiator on agriculture. Julius Katz became Hills's deputy and took an active role in the agricultural talks. On the EC side, Frans Andriessen moved from being Agricultural Commissioner to External Affairs, and therefore took over from Willy de Clercq as chief trade negotiator. The new Agricultural Commissioner was Ray MacSharry. The new teams showed a modicum of flexibility and, anxious to get the round back on track, moved quickly to a resolution of the agricultural issue.

The April Agreement

The key to breaking the impasse on agriculture was the agreement on the wording of the objective of the negotiations. In place of the aim of eliminating support, the April Agreement called for 'substantial progressive reductions in agricultural support and protection sustained over an agreed period of time', in order to prevent 'distortions in world agricultural markets'. This form of words allowed both sides to declare victory: the EC could rightly claim that it had not agreed to a total removal of price support, and the US could point to the logic of substantial and progressive reductions as eventually leading to the removal of trade distortions. The April Agreement resolved none of the issues, but it allowed talks to resume and gave negotiators a focus and a timetable. Countries had until the end of 1989 to present 'detailed proposals for the achievement of the long-term objective(s)' of the negotiations, and were asked to consider the range of modalities already on the table for addressing areas such as import access, export competition and domestic subsidies – as well as sanitary and phytosanitary trade impediments and the role of developing countries. In addition, a freeze on support price levels was agreed, the first time such an undertaking had ever been attempted. As world prices had firmed considerably from their 1987 lows, and as country after country was trying to contain spending on farm policies, the freeze was probably not an undue hardship to most governments.[30]

As a lead-up to the 'comprehensive proposals' called for in the April Agreement, other papers were produced by the main protagonists during 1989, giving the views of individual countries on specific aspects of the negotiations. Among the most important of these were the paper elaborating on the US proposal for tariffication[31] and that from the EC explaining their development of the AMS, the Support Measurement Unit (SMU). These ideas were then embodied in papers which addressed all the issues under discussion.

The major protagonists duly produced their comprehensive proposals at the end of 1989, and the second major set of confrontations in the negotiations began. The positions of both the EC and the US had changed dramatically since the first proposals in 1987, and significant shifts had also taken place in those of the Cairns Group and Japan. Unfortunately, there was little evidence that these shifts had expanded the common ground between the positions. The US, for instance, abandoned its reliance on an overall support measure just as the EC became its champion. The changes seemed to reflect the evolution of thinking in the capitals, coupled with a deep suspicion of the motives of negotiating partners, rather than an attempt to forge a compromise among the positions. The real negotiations were not in fact to begin until much later.

The US Comprehensive Proposal

The US 'Comprehensive Proposal' of October 1989 laid out a detailed programme for the negotiations in agriculture. The most important departure from the July 1987 paper was the emphasis on negotiating new rules ('disciplines') aimed at four different areas of policy: import access, export competition, internal support, and sanitary and phytosanitary measures. The US called for reform in all four areas. Non-tariff import barriers would after a transition period be converted to tariffs and then ultimately reduced 'to zero or low levels', allowing all waivers and derogations to be eliminated along with Article XI: 2(c), the clause that allows quantitative import restrictions as a back-up for domestic supply control (GATT, 1989b). The conversion to tariffs of non-tariff import barriers became known by the ungainly word 'tariffication'. All existing tariffs would be bound and then reduced over a ten-year period. Existing non-tariff barriers would for the transition period be replaced by tariff-rate quotas (TRQ), set in relation to current import levels or negotiated where those levels were deemed to be low. Only tariffs would be allowed on above-quota trade. The initial TRQ would be expanded and the above-quota tariffs reduced over the ten-year period. This would 'permit an orderly transition from the extremely high levels of import protection provided by some current non-tariff barriers to a tariff-based import regime'. Export subsidies would be phased out altogether, along with export prohibitions, over a five-year period. Only bona fide food aid would be exempted, but to avoid abuse of this exception 'new rules may need to be developed to govern the granting of food aid'. To curb the trade-distorting aspects of internal support, domestic subsidies would be categorized. Those deemed most injurious to trade would be phased out over a ten-year period; those that were deemed to be non-trade distorting would be permitted; and those on the borderline would be disciplined 'to prevent their use in ways that would nullify or impair concessions or cause serious prejudice or material injury' to others. This last category would also be subject to negotiated reductions. In the two years since the first salvos were fired, the US had moved from advocating a general phase-out of trade-distorting support, based on an aggregate measurement of support (AMS), to seeking actions on specific policies such as non-tariff barriers and export and domestic subsidies. With a zero AMS target for every country and every commodity, as in the 1987 US proposal, the types of policies that countries used in the transition period would have been largely immaterial. With partial liberalization rather than a full withdrawal from price supports as the new, more modest aim of the US, the method of support once again seemed significant.

The US 'Comprehensive Proposal' in effect laid down a direct challenge to the EC to abandon its support system, the Common Agricultural Policy. The variable levy would have to be turned into a fixed tariff, export subsidies would be eliminated, and domestic subsidies tied to domestic production, such as oil-seed deficiency payments, would be phased out. Policies that would be allowed would include research and extension, domestic food subsidies, food aid, and 'decoupled' income payments to producers. The EC market management system would have been effectively dismantled. The cool reception by the EC was not entirely unexpected.

The EC Comprehensive Proposal

Meanwhile, the EC had also been revising its position, abandoning the demand for short-term action (market prices were much higher in 1989 than in 1987) and warming to the notion of a steady reduction in overall support, though still not contemplating its elimination. The EC's 'Global Proposal' of December 1989 stated plainly that 'any negotiation that focused in priority on frontier measures would in no way contribute, in contrast to what a superficial analysis might suggest, to an improvement of trade' (GATT, 1989c). For that reason, the Community believed 'that the commitments to be taken to reduce support and protection must be made in terms of an aggregate measure of support'. The use of an AMS had moved back into the spotlight. On tariffication, the EC was still sceptical. The paper acknowledged that 'the means by which support and external protection are ensured is . . . a source of serious difficulties' for world markets, but then went on to define a set of trade problems which arose from uneven levels of protection for different commodities.[32] This somewhat idiosyncratic definition allowed the EC to proclaim that 'tariffication does not provide a reasonable or convincing solution to these types of problems' (GATT, 1989c, p. 5). As if to emphasize the underlying differences between their position and those of the US and the Cairns Group, the EC paper added the observation that 'basing protection exclusively on customs tariffs and envisaging, after a transition period, the reduction of these tariffs to zero or a very low level would lead to trade in agricultural products on a totally free and chaotic basis.' However, the EC paper was 'prepared to consider including elements of tariffication in the rules of external protection given that the problem of rebalancing can be solved in the context of tariffication'. In addition, the 'fixed component' of border protection would itself be 'completed by a corrective factor' to take into account exchange rate variations and world market fluctuations.[33] In addition, the Community suggested that 'deficiency payment[s] would be treated in the same way and converted into tariffs', presumably so that the US would have to accept similar disciplines for its own domestic programmes.[34]

The US noted with some satisfaction that the EC could live with tariffication, even though subject to safeguards, but rejected strongly the notion of 'rebalancing' protection. In particular, any attempt to increase trade barriers on feed ingredients that had been the beneficiaries of the low, bound tariffs accepted by the EC in the Dillon Round agreement was ruled out in no uncertain terms.

Other Comprehensive Proposals

The Cairns Group tabled its own Comprehensive Proposal in December 1989 (GATT, 1989d). It envisaged a ten-year reform process applying 'to all measures affecting agricultural trade, directly or indirectly'. The proposal embraced the notion of tariffication of non-tariff import barriers, subject to a maximum ad valorem level and agreed reduction rates, and endorsed the idea of 'global tariff quotas' as a temporary device to ensure the continuation of current access. The tariff quotas would be removed when tariffs were finally bound. The Cairns Group supported strongly the eventual elimination of export subsidies. Current levels of export subsidies, per unit and total outlay, would be frozen and progressively phased out. On domestic subsidies the Cairns Group proposal was inclined towards prohibiting the most trade-distorting types of instrument, and argued for cuts in both a product-specific AMS and in producer support prices to implement the agreement. However, the Cairns Group added that countries could be accorded flexibility in choosing the policy mix to achieve the reductions. Support not linked to production or trade would be exempted from the reform commitments. The Japanese position, like that of the EC, had also undergone modification, with somewhat greater acceptance of the notion of action on domestic subsidies, but the Japanese paper continued to argue that non-trade concerns should be allowed to put some domestic programmes beyond the reach of international trade rules (GATT, 1989e).

Negotiators continued the process of exploring and elaborating these diverse positions in the first half of 1990. Though the amount of common ground had considerably expanded, much remained to be done before an agreement could be forged. In particular, negotiators would eventually have to focus on one single document, rather than continue to discuss the proposals of individual countries.

The Chairman's Draft and the Final Offers

The first attempt at such a unifying draft was that put together by the Chairman of the Agricultural Negotiating Group, Aart de Zeeuw, in July 1990. The 'Chairman's Draft' called for specific commitments to be made on import access, export competition, and domestic subsidies. On import access (border protection) the draft agreement called for 'conversion of all border measures . . . into tariff equivalents' based on 'the existing gap between external and internal prices'; a binding of these new (and existing) tariffs; maintenance of current access through tariff quotas; and 'in the absence of significant imports, establishment of a minimum level of access' based on a proportion of consumption of the product concerned (GATT, 1990a). The new and existing tariffs should be 'substantially and progressively reduced' over a number of years. The level of reduction was left unspecified. The import access commitments were to be tabled in the form of country lists. The Draft called for 'direct budgetary assistance to exports, other payments on products exported and other forms of export assistance' to be 'substantially and progressively reduced', and offered a menu of choices as to how this could be done. No rate of reduction or timetable was offered, but export subsidies were to be reduced faster than import barriers or domestic subsidies. On internal support, the Draft also called for reductions 'so as to minimize trade distortion and increase the market orientation of production'. Included in such support were to be market price support, direct payments, and input and marketing cost reducing programmes. Country lists were to be drawn up which would establish the baseline AMS from which reductions would made. The Chairman's Draft therefore accepted the structure suggested by the US and the Cairns Group, including their emphasis on export subsidies, but left the pace of protection and support reduction to be negotiated – thereby in effect allowing the Community and Japan to impose some restraints on the process of liberalizing domestic policies. The Draft called for the national 'offers' to be submitted by 1 October 1990, including numerical targets for support reduction and access expansion.

The De Zeeuw text was a milestone in the Round, and indeed in agricultural trade relations in general. It defined the areas of negotiation and the options in a clear, concise manner and made apparent the steps needed to get agreement by the date of the Final Meeting in December 1990. Its framework survived in the Final Act of the Uruguay Round, and so can be considered the first version of the ultimate agreement. Most negotiating parties took the Chairman's Draft as a viable document on which to focus. The EC, however, had severe reservations, in particular regarding the notion of discussing separate limits on export subsidies. The EC was in effect isolated in its position, though a number of countries such as Korea and Japan for rather different reasons may have been silently hoping that the EC would stay firm in opposing the Draft. The US made a concerted effort to have the Draft accepted as a basis for detailed negotiations. No political channels were left unused. A particularly visible but unsuccessful attempt was made by President Bush at the Houston Summit of the G7 countries in July 1990 to make the Chairman's Draft the basis for negotiations to end the Round. Although all participants agreed that it could be a base from which to 'intensify' negotiations, the EC did not accept its structure as having been agreed.

When they came to present their 'final' offers with concrete numerical proposals, the main protagonists were still far apart. Indeed these 'final' positions proved to be ultimately irreconcilable. The US paper of October 1990 took the Chairman's Draft as a framework, modifying and extending it somewhat and adding the quantitative dimension (GATT, 1990b). It called for a 75 per cent cut in domestic support (as measured by a commodity-specific AMS) and in tariffs (including those newly formed by conversion of non-tariff import barriers), and a 90 per cent cut in export subsidies, all over a period of ten years. In addition, a sector-wide AMS would be subject to reduction commitments of 30 per cent over the implementation period.[35] At the end of the ten-year implementation period, countries were to review the agreement and evaluate the appropriateness of further reductions. The tariff-rate quotas guaranteeing access were to be removed after ten years, and 'in no case, however, shall any tariff above 50 percent' be maintained beyond the implementation period (GATT, 1990b, p. 6). The US offer also elaborated on the issue of safeguards against import surges and low prices in the case of markets subject to tariffication, giving countries a choice between a quantity trigger and a price trigger to activate temporary relief measures. The Cairns Group in their offer broadly agreed with the timetable and depth of cuts spelled out in the US position, though Canada expressed some reservations on the import access issue.

The EC, by contrast, chose to stick to its previous position that separate commitments on import access, export competition, and domestic subsidies were both unnecessary and undesirable. Its own paper, tabled in November 1990 only after seven acrimonious debates among EC ministers, went only a little way toward meeting the US and Cairns Group positions. Instead of reductions of 75–90 per cent over ten years in levels of support and protection, the EC offered overall support reduction as measured by an AMS of 30 per cent over five years (GATT, 1990c). It agreed to 'the tariffication of certain border measures and a concomitant reduction of the fixed component resulting therefrom, together with a corrective factor' but insisted that tariffication be 'subject to rebalancing'. This element of rebalancing was defined (for the EC) as a tariff based on the cereal levies, on oil-seeds, protein crops, corn gluten feed and other non-grain feed ingredients, with tariff-rate quotas for quantities equivalent to existing imports which could enter at 6 to 12 per cent tariffs. On export subsidies, the EC predicted that 'the proposed reduction of support and protection will lead to a considerable reduction of export subsidies', and stated that it was 'ready to quantify the results [for export subsidies] flowing from the reduction in internal support'. The EC also agreed not to introduce 'export subsidies for commodities for which they have not been applied in the past'. It was clear that, although it had clarified some positions and 'intensified' the discussion, the Chairman's Draft had failed to become the unifying document needed to speed agreement at the forthcoming meeting in Brussels. The EC was still not on the same page.

The Brussels Meeting

The lack of an agreement on the framework continued until the opening of the Brussels meeting of the Trade Negotiation Committee (TNC) in December 1990. In place of a single document, the negotiators had before them the various national positions and some 'clarifications' elicited by the Secretariat. The task of finding an agreeable compromise fell to the Swedish Minister of Agriculture Mats Hellström, as chairman of the agricultural negotiating group at the ministerial level. In an attempt to get an agreement, Hellström circulated a 'non-paper' which suggested a 30 per cent reduction in internal support, border protection and export subsidies, from a more recent base than proposed in the EC position.[36] All delegations were required to respond to the proposals privately to Hellström. The draft found broad support from the US and the Cairns Group, and from a number of other delegations. The EC Commission also submitted a somewhat cautious but broadly positive response. For a few hours it appeared there was a breakthrough. The US delegation quickly began to move in other areas, above all services, and other delegations also began to prepare for final agreement. However, later the same day France and Ireland told the EC Commission that in its response to the Hellström draft it had exceeded its negotiating mandate. When the agricultural negotiating group got together for its next meeting that night (December 6), EC Commissioner MacSharry therefore partly withdrew from the Hellström proposal. The atmosphere was electric, and it soon became clear that no compromise could be reached.

The other negotiating groups waited for the word that a breakthrough had been achieved in agriculture. However, when the expectations that the Hellström formula would be acceptable were dashed, the US and the Cairns Group countries within five minutes withdrew their delegations from the other negotiating groups. It became clear that the Brussels meeting had failed and the Round had to be suspended. Amid confusion and consternation the Brussels TNC meeting was adjourned.

Phase Three: From CAP Reform to Marrakesh

The US and the Cairns Group made it clear in the aftermath of the Brussels TNC meeting that negotiations could only start up again if the EC indicated some movement on agriculture. There followed a series of bilateral discussions between the EC and the US, and a series of visits to national capitals by the GATT Director-General Arthur Dunkel aimed at keeping the Round alive. Dunkel's 'shuttle diplomacy' paid off. In late February the EC indicated that it was willing to change its position on certain key issues, and – most importantly – that it was now willing to negotiate separate commitments on import access, export competition and domestic support. The Round was revived yet again.

The breakthrough had come from within the EC bureaucracy. Radical changes in the CAP were under discussion in Brussels which would make the GATT negotiations less of a threat to the operation of Europe's common agricultural policy. The Community had resisted for much of the GATT round the notion that it must change its policy instruments. On the very day of the breakdown of the GATT talks, a document (also called a 'non-paper') was circulating in Brussels in which Ray MacSharry, the EC Commissioner for Agriculture, proposed a radical policy departure. The support price for cereals, oil-seeds, and pulses was to be cut by a significant amount, later quantified at 35 per cent, to be offset by hectarage compensation payments to farmers. Dairy and beef prices were also to be cut, with compensation paid to these producers. Despite vigorous denials by the European Commission at the time, the MacSharry CAP reform proposals were in large part an outcome of pressures from trading partners in the GATT. [37]

The implication for the GATT round was clear. The Community could, if it introduced such reforms, live with a requirement for a cut in support of Hellström proportions. The market price cuts would stimulate consumption, and, along with some reduction in output from a set-aside programme, this would enable the Community to meet the targets for reduction in export subsidies. Furthermore, if the 'compensation payments' were to be excluded from the calculation of the AMS, the EC would be well within the 30 per cent reduction target for internal support without making further cuts in support prices. In the light of this change in position the EC reluctantly removed its objection to negotiations along the lines accepted by others, and talks began in earnest.

The Dunkel Draft

The change in the EC position rejuvenated the talks. Arthur Dunkel once again undertook a tour of the capitals to find out what might now be possible. Technical discussions proceeded at a faster rate and with more purpose. The culmination of this burst of activity was the Draft Final Act (usually called the Dunkel Draft) of December 1991, which brought the provisional texts from the various negotiating groups together into one document (GATT, 1991). In other areas, the text had largely been agreed among the negotiators by the end of the Brussels meeting: in the case of agriculture, Dunkel took it upon himself to define the compromise position and to sell it to the negotiating parties.

The Dunkel Draft introduced few new ideas but developed in a more concrete form those first found in the De Zeeuw Draft and later in the Hellström non-paper. The conversion of non-tariff import barriers into fixed tariffs, the constraining of export subsidies, and the categorization of domestic subsidies as 'amber' or 'green' depending upon their trade-distorting nature were all included. The concept of a 'red' box had by this stage been dropped. No domestic policy was to be disallowed, but all except the green box measures were to be included in the AMS and be subject to reduction. In addition, the Dunkel Draft suggested a timetable and quantitative targets for the 'reform programme' for agriculture to reduce the protective effect of existing policies by scaling back support. Both new and existing tariffs would be reduced by 36 per cent over six years (based on the period 1986–8); export subsidies would be scaled down by 36 per cent in expenditure and 24 per cent in volume (relative to 1986–90); and domestic support measures in the amber box would be reduced by 20 per cent in six years (from their 1986–8 levels). In addition, a new accord on health and phytosanitary trade restrictions was to be included in the GATT package, which would improve the transparency of national regulations in this area and make the settlement of disputes easier. A timetable for completion of the talks was reintroduced, with an initial deadline for broad acceptance of the Draft Final Act of 20 January 1992, to be followed by a submission of draft schedules ('offers') indicating what individual countries were prepared to do if the Draft were to be agreed.[38] A final date of 15 April 1992 was set for the end of the Round, to allow the completion of the negotiations before the expiry of the US 'fast-track' authorization. Implementation of the agreement was to begin in January 1993.

The Dunkel Draft was warmly embraced by the US, the Cairns Group and most of the other participants in the negotiations. The EC was less than happy with the timetable and depth of cuts proposed. Also, the EC's demand for 'rebalancing' had not been met; the AMS to be used to monitor support reduction had not included 'credit' for supply control; and the compensation payments being discussed in the context of CAP reform appeared to be included in the amber box of objectionable policies; and the EC was being asked to make a commitment to a cut in the volume of commodities eligible for export subsidies. The EC responded in guarded terms by submitting its list of base period tariff equivalents, AMS levels, and export subsidies, but it did not offer substantive commitments for reduction.

The negotiations proceeded on a technical level throughout the spring and summer of 1992, with little visible progress. The deadline of April 1992 passed, and the US Administration had to request from Congress yet another extension of 'fast-track' negotiating authority (for six months) to keep the negotiations alive. The EC grappled internally with the MacSharry reforms for the cereal and oil-seed sectors, and to the surprise of many adopted the new policy in June 1992. The US was occupied with a presidential election which concentrated on domestic economic issues. Negotiations on the NAFTA were also drawing to a close, and that topic attracted more immediate interest than the GATT round. Meanwhile, a further agricultural issue appeared on the agenda to complicate the task of forging a compromise. The new item was the US complaint in the GATT about the EC's oil-seed regime, and in particular the subsidies paid to processors to allow them to use more expensive domestic oil-seeds in place of those imported from the US and other countries. At the risk of breaking the chronological thread, the parallel development of the oil-seeds dispute is worthy of explication.

The Oil-seed Dispute

The historical origin of the oil-seed dispute dates back to the Dillon Round when the EC had bound zero tariffs on oil-seeds and protein feeds, thereby 'paying' the US for the unbinding of tariffs on central CAP products, including cereals (see Chapter 3). As a consequence, the EC could not establish for oil-seeds the same type of domestic price support as it granted to cereals and other crops. In order still to provide some production incentives and income support to oil-seed growers, the EC introduced a domestic subsidy. For administrative ease, the subsidy was paid to crushers on the amount of EC-produced oil-seeds purchased by them. The level of subsidy was calculated so as to compensate crushers for the difference between the EC target price and the world price for oil-seeds. Over time, EC oil-seed production had grown significantly, and US exports of oil-seeds and protein feeds to the EC had declined. The US complaint had two parts. First, the US claimed that the EC subsidy violated the national treatment obligation of the GATT (Article III:4) because it made it more profitable for crushers to use EC rather than imported oil-seeds. Second, the US maintained that the resulting production incentives to EC oil-seed growers had caused 'nullification and impairment' of the benefit the US could expect to reap from the EC tariff binding.

The panel essentially agreed with both claims. It found that the EC regulations were such that the subsidy paid to crushers was sometimes more than necessary to compensate for the difference between the target and world price, resulting in less favourable treatment of imported oil-seeds than that of EC oil-seeds, which was inconsistent with Article III:4. On nullification and impairment, the panel reasoned that the subsidy protected EC oil-seed growers completely from the movement of world prices, thereby eliminating price competition between domestic and imported oil-seeds. As improved price competition was one of the primary advantages of a tariff binding, the subsidy impaired benefits accruing to the US. The EC noted its reservations to this rather firm panel ruling, but did not block its adoption by the GATT Council. Indeed, the EC adjusted its oil-seed policy in December 1991 such that subsidies were paid directly to oil-seed growers, on a per hectare basis. The US conceded that this policy change had healed the violation of the national treatment principle, but maintained that nullification and impairment still occurred because even under the new regime EC producers were largely shielded from world market price changes, since the EC had designed its per hectare payments scheme such that aid was higher when world prices were lower, and vice versa. On a US request the same panel was reconvened and made a second ruling, again essentially agreeing with the claims made by the US. This time the EC blocked adoption of the panel report, but requested and obtained authority to renegotiate the original tariff binding (under GATT Article XXVIII:4). After heated and unsuccessful negotiations, involving threats of retaliation and counter retaliation, a settlement was finally found in November 1992, as part of the Blair House Accord which opened the way towards agreement in the overall Uruguay Round negotiations on agriculture.

The oil-seeds case was important on several counts. It showed that GATT panels were able to issue firm rulings, and in a reasonably short time. It dealt with a matter of huge commercial significance in agricultural trade. It was the only case in which a central element of a market regime under the CAP was successfully challenged in the GATT, though the market regime concerned was somewhat of an abnormality in the EC because it did not rely on variable levies. The EC essentially accepted the (first) panel ruling and changed its policy accordingly, and the final settlement in the case was part of a deal which provided a breakthrough in the Uruguay Round.[39]

The Blair House Accord

It took the EC and the US almost a year of further negotiations after the Dunkel Draft to reach an agreement on the terms of the agricultural component of the Uruguay Round. This momentous event occurred in late November 1992 at a bilateral negotiating session at Blair House, in Washington. The Blair House Accord represented the political culmination of the six-year confrontation between the US and the EC. The high point of the drama, however, was three weeks earlier, when the US and EC officials met in a Chicago hotel to try to bring the Round to an end (and to settle the dispute over EC oil-seed subsidies) before the US election.[40] George Bush considered a successful GATT round a political benefit, and had been pressing European leaders for flexibility on the final terms. The EC delegation was led by Ray MacSharry, the Agricultural Commissioner, operating with the blessing of the Trade Commissioner, Frans Andriessen. On the US side sat the Secretary of Agriculture, Edward Madigan. A deal was all but struck by the election date (Madigan left the talks to cast his vote), but MacSharry was informed by phone from Commission President Delors in Brussels that he had exceeded the negotiating mandate given by the Council of Ministers to the Commission. MacSharry returned to Brussels and resigned his role in the agricultural talks, and only resumed negotiations after being assured that he had the support of his fellow Commissioners.[41] Meanwhile, the US announced punitive tariffs on $300 million worth of EC agricultural exports (mostly white wine from France and Italy) in retaliation for the impaired access to the EC oil-seed market. The tariffs were due to go into effect if a settlement were not reached within 30 days.

After these histrionics, the meeting at Blair House was anticlimactic. President Bush, on his way out of office, wanted to leave a legacy, even if the goal of a conclusion to the Round on his 'watch' had proved elusive. Agreement was reached on 20 November 1992 between the US and the EC on a package which included the agricultural component of the Uruguay Round and the oil-seed dispute which had come to be entwined with the Round.

The Blair House Accord modified, and weakened, the terms of the Dunkel Draft agreement on agriculture. At the insistence of the EC, the export subsidy quantity reduction was scaled back to 21 per cent (from 24 per cent) and the AMS commitment on domestic support was modified to apply to an aggregate of all commodities rather than to each commodity. It was also agreed that both the EC and the US direct payments programmes for cereals would be sheltered in a 'blue box', as policies that, if not truly decoupled, at least had acreage set-aside provisions associated with the payments. A 'peace clause' was added, to give some degree of shelter from GATT challenges to policies that were consistent with the aims of the Accord, and a gentlemen's agreement was reached to reconsider the issue of non-grain feeds if imports into the EC were to rise too strongly. Lastly, a solution to the long-running oil-seeds conflict was wrapped in the package. This enabled the US to claim that it had limited the expansion of oil-seed production in Europe by agreeing to a maximum hectarage under oil crops, and the EC to claim that it had achieved recognition of the legitimacy of the CAP reform programme.

The Final Act

The Blair House agreement was a bilateral deal. To form the basis for the Final Act of the Uruguay Round it had then to be accepted by the other negotiators, notably the Cairns Group. This caused some soul-searching among the other exporters. The main elements of the package from the de Zeeuw and Dunkel drafts were intact: the Blair House Accord was in effect an agreement between the US and the EC as to how they would interpret and implement the provisions of the Dunkel text. But the bilateral deal had watered down some of the impact of that text by ensuring that the cereals support payments of the US and EC escaped direct challenge, by weakening the export subsidy quantity provision, and by rendering the constraint on internal support less effective by applying it not to each specific product but to an aggregate across all commodities.

Even among the parties to the Blair House deal there was more to be done before the agreement could be concluded. The French government, supported by vocal rural interest groups and parliamentary pressure, wanted the Commission to renegotiate parts of the agreement. At issue was whether the Commission had gone beyond its mandate in reaching the deal. Underlying the concern was a fear that to abide by the deal the EC would have to extend and deepen the 1992 reforms, perhaps by further set-asides. The elaborate compromise of Blair House was therefore refined in last-minute negotiations between the EC and the US in Geneva in December 1993. The nature of the final deal (sometimes called Blair House II) was to delay the immediate impact of some of the export subsidy cuts implicit in the Blair House reductions.[42] Another element which was agreed in these last-minute negotiations was a commitment by the EC not to use the full bound tariffs on cereals if that would result in import prices higher than 155 per cent of the EC intervention price. A somewhat nebulous formula was agreed, according to which countries would consult annually with respect to their 'participation in the normal growth of world trade in agricultural products within the framework of the commitments on export subsidies'.[43]

The time between the end of the negotiations in Geneva and the meeting scheduled for April in Marrakesh at which the Agreement was to be signed was devoted to the 'clarification and verification' of the individual schedules. It was these schedules that contained the details of the implementation of the agricultural (and tariff reduction) parts of the Agreement. This proved to be a more significant step than had been anticipated. Countries had guidance in the 'Modalities' paper (GATT, 1993) on the way in which tariffs were to be calculated and the amount by which they had to be reduced, but the process of tariffication and the schedule of reductions were not adequately supervised. Once one country had engaged in the 'creative' interpretation of the agreed guidelines then it was difficult for others to resist following suit. The end result was considerably less liberalization of trade than had been intended by the negotiators collectively when agreeing to the package in December.

3   THE SIGNIFICANCE OF THE AGRICULTURAL NEGOTIATIONS

The Uruguay Round negotiations on agriculture, in particular from July 1987 to December 1990, provided an ongoing, public symposium on the issues plaguing the world markets for temperate zone commodities and the impact of domestic farm policies.[44] Its conclusion left a new set of rules to guide agricultural trade, along with detailed schedules of tariff rates, agreed reductions in export subsidies and limits on domestic subsidies. The significance of the agricultural negotiations lies both in the debate itself, with the evolution of ideas and perceptions, and in the eventual outcome. The legacy of the Round is likely to be enduring: the discussions took place at a time of rapid evolution of economic policies and political structures, and the outcome of the negotiations both incorporates and reinforces many of these momentous changes.

The agricultural trade debate was not confined to agricultural officials, though the technical and detailed nature of the proposals ensured that these officials and their counterparts in the private sector kept effective control. Perhaps for the first time, issues of agriculture were taken seriously by those with no interest in agriculture as such. The issues had become too important to be left to the agricultural interests alone. Not only was it important to get an agreement to secure the overall promise of the Round, but countries seemed willing to try for a constructive agreement that would change permanently the conditions of agricultural trade. This section of the chapter looks at some of the contributions that the Uruguay Round discussions themselves have made to the beginning of the long, slow process of improving the conditions for agricultural trade.

The Negotiating Framework

It was clear to all parties, even those who would have preferred it otherwise, that a new approach was needed for the conduct of agricultural trade negotiations. Everyone recognized that a significant reform of agricultural trade rules and practices was long overdue.[45] It also became widely accepted early in the process that the national agricultural price support policies pursued by the major trading countries were at the root of the problems in agricultural trade, and therefore that these policies, in particular those of the major industrialized countries, must be modified before the trade situation could be noticeably improved.[46] What was less clear was the nature of the rule changes, the extent and speed of the process of reform of price support policies, and the extent to which the international community, through the GATT, should mandate such reform. The search for a framework for negotiations was therefore a search for the most appropriate way of incorporating constraints on domestic policies into international trade rules.[47]

Rules and Reductions

The path to such domestic policy constraints could have taken a number of directions. It could have taken an indirect course through a change in the trade rules, either banning the use of particular trade instruments, such as non-tariff import restrictions or export subsidies, in order to take away those instruments that were most responsible for the 'export' of the problem of high internal price supports. This would have targeted those countries that made heaviest use of the problematic instruments of protection. A similar effect could also have been accomplished by scaling down the level of those trade instruments deemed to be objectionable, through reductions in tariffs and relaxation of quotas and limits on the levels of export subsidies. This would indirectly put pressure on domestic price policies, without the need to negotiate directly on their level. Alternatively, countries could opt for an agreement that mandated a change of farm policy instruments operating within the border. A switch from deficiency payments, for instance, to decoupled income payments would have been possible, so that the same support could be given in a less trade-disruptive manner. This would have meant intruding far into domestic policy decisions. Finally, countries could have agreed to a scaling down of the overall level of support given by farm policies, so as to reduce the amount of potential trade disruption.[48]

The nature of the approach chosen was understood to have profound implications for the outcome of negotiations. The majority of the five years between Punta del Este and the Dunkel Draft was in fact spent defining the approach. The easy way out would have been to stick to discussing trade instruments and the level of trade barriers. Negotiating on trade instruments and the level of trade barriers is clearly easier at an international level than negotiating domestic policy reform or price support levels.[49] But the consensus, at least among the exporters, was that this by itself was not enough. Trade barriers, such as the EC's variable levy and the quantitative import restrictions in Canada, were so much a part of the internal agricultural policies of the industrial countries that meaningful changes could not have been made without radical changes in those policies. Domestic policies had to be on the bargaining table. However, negotiating directly on the substance of domestic policy was likely to be too difficult.[50] Farm groups in the US were insistent that 'farm policy is made in Washington, not Geneva'.

The agricultural negotiations from 1987 to 1991 constituted a search for the appropriate balance between these four approaches. All four found their way into the final package for agriculture. The trade 'rules' approach, the standard GATT technique for correcting problems with the operation of the trade system, was initially passed over for agriculture in favour of that of negotiating overall support reductions, but it reappeared in mid-negotiations (in the form of proposals for tarification and a ban on export subsidies) as a more direct way to ensure improvements in agricultural trade. The trade barrier 'reductions' approach was employed in the tariff-cutting and access-increasing elements of the agricultural package, as well as in the phasing down of export subsidies. The application of the 'rules' approach to domestic policy found its manifestation in the 'green box' classification of domestic subsidies. Though not prohibiting particular types of domestic subsidy, the green-box criteria changed the attractiveness of particular policy instruments and sheltered some programmes from external challenge. The schedule of cuts in overall support mandated in the agricultural agreement exemplifies the 'reduction' approach applied to domestic policy instruments. In some ways this is the most novel, though somewhat weak, part of the agreement. But the fact that it was, in the end, used at all could have significant implications for future trade talks in agriculture. Domestic support levels, once on the international negotiating table, are unlikely to be able to escape international controls.

The tensions between negotiating on domestic programmes or on trade instruments, and between rule changes and level reductions, were at the heart of the ongoing dialogue between the US and the EC. The US had the internal agricultural policy of the EC clearly in its sights. The problems posed for other exporters by the CAP were that domestic support price levels were too high, leading to restricted access and to exported surpluses, and that the instruments used to maintain these internal prices, variable levies, and export subsidies, isolated and insulated the European market and disrupted world trade. The Cairns Group concurred with this assessment, but added a complaint of its own that the US reaction to the EC policy was itself disruptive to other exporters. If all price support policies were to be removed, the problem would disappear (along with the CAP). If support levels were merely to be reduced, the problem of the instruments would still remain. In agricultural trade demonology, the Community had been able to avoid external discipline on the CAP largely by the use of the variable levy. To turn this unloved instrument into a tariff offered a real prize. However, converting the variable levy into a bound tariff would still leave open the issue of export subsidies. These also needed to be circumscribed with rules and subjected to reduction. And to prevent the EC from instituting open-ended deficiency payments in place of protection at the border, rules for domestic subsidies were needed, along with some ceiling on overall support. Thus, once the 'zero option' was dead, negotiations on instrument-specific rules seemed the next best strategy.

The EC had difficulties with the notion of tightening the rules on trade instruments for precisely the reasons the US and the Cairns Group wanted them. The EC made more extensive use of border measures to implement its domestic price support policies than did other countries. Any attack on non-tariff import barriers and on export subsidies was a frontal assault on the CAP. By contrast, a negotiation on the overall level of support, cutting down such support over a period of years, was not only more acceptable as it did not require painful changes in policy instruments but also more desirable as a way of putting external constraints on the tendency of CAP spending to get out of hand.[51] The compromise which emerged in the Dunkel Draft was only possible after the EC had modified the CAP to make it less reliant on trade instruments.

The Aggregate Measurement of Support

The use of an aggregate measurement of support promised to resolve the dilemma facing all agricultural trade negotiations, that the national programmes that cause the trade problems are not amenable to international control. The hope offered by the AMS approach was that one could negotiate the magnitude of the trade impact of domestic policies without discussing their detailed implementation. The key was first to agree on the terms of such a measure, and then to use it to define and monitor support reduction commitments. The preferred version of the AMS that emerged in the 1987 negotiating proposals was the Producer Subsidy Equivalent (PSE) made familiar by the OECD Secretariat Trade Mandate study (see Chapter 6). In addition to defining the type of measure, one must also decide which policies to try to capture or isolate: i.e. one has to attempt to define which set of domestic programmes is to be considered as disruptive to trade. This led in turn to the definition of rules on the acceptability of various instruments of domestic policy.[52]

The benefits of using an AMS were clear to trade negotiators. If one could perform the trick of binding support levels in an international negotiation then the task of scaling down the policies themselves could be left to governments as a part of internal policy. Some degree of monitoring would be needed, but no international 'micro-management' of national policy changes would be called for. Economists warned, however, that some caution was needed in interpreting an AMS as an exact proxy for the trade effect of domestic policies.[53] Some programmes, for instance, tied support payments to supply control. These presumably had less impact on trade than did unconstrained policies. Other policies transferred resources to producers in a way that affected consumption (and hence trade), while others involved more direct subventions from the treasury. The former were likely to be more objectionable to trade partners. Moreover, trade negotiators themselves began to be concerned about the desirability of agreeing to AMS reductions when such measures were themselves influenced by changes in world prices and exchange rates.

The debate on the use of the AMS began to wane in the summer of 1990. Until that time the notion of scaling down total agricultural support was still an option for the negotiating parties. The de Zeeuw draft had included it in the realm of possible modalities, without making it central to the process. Its role was seen more as a monitoring device than as a vehicle for binding commitments. The increasing enthusiasm of the EC for using such a measure was matched by the growing wariness of the US. In the end, the AMS survived only as a measure for the control of internal subsidies. This measure was to be reduced by a scant 20 per cent over the six-year reform period, with a lower reduction for developing countries. Moreover, the restriction applies to the total support across all commodities, thus making it virtually useless as a constraint on individual commodity support programmes. The weakening of the constraint offered by the 'global' AMS in turn reduced its meaningfulness as an instrument of reform. Rather than leading the way by forcing reform on reluctant countries, it acts as a back stop in case those unwilling to reform would try a massive switch of support to policy instruments not otherwise captured in the new rules. The reduction target arguably should have been at least equivalent to that set for import barriers and export subsidy expenditures, but by the end of the negotiations the general feeling was that this part of the agreement in any case had few teeth. There was therefore little point in attempting to negotiate significant cuts in the AMS.

New Rules for Old Problems

The Uruguay Round profoundly changed the rules governing agricultural trade. The nature of the rule changes is without doubt the most interesting aspect of the outcome of the agricultural talks.[54] These new rules and procedures emerged during the progression of the Round.

Few at the outset would have believed it possible to make such changes to a part of the trade system which had for so long resisted reform. The most dramatic rule changes concern the conversion of non-tariff import barriers into bound tariffs, the prohibition on new export subsidies and the categorization of policy instruments into amber and green boxes. The quantified constraint on existing export subsidies was a decisive departure from the unworkable concept of 'equitable share', and the curb on internal support levels is a novelty of potential importance though with a less immediate impact. Each rule change and constraint reinforces the others: the totality is a firm control of the overall level of support and protection available to agriculture.

Tariffication

The use of non-tariff trade barriers has always been a central concern of the GATT. The General Agreement provides for the use of tariffs as the only border instrument, subject to limited exceptions, and the GATT is well versed in the lowering of tariffs, the classic form of trade negotiation. The Uruguay Round agricultural negotiations mandated the conversion of non-tariff trade barriers to tariffs and included the scaling down of those tariffs over a six-year period. The exemption from the tariffs-only rule that was made for some rice producers is intended to be of a strictly temporary nature.

Tariffication was not universally popular. The EC, in particular, was reluctant to move to 'pure' tariffs, which would allow domestic prices to rise and fall with world prices. The EC had suggested a price-trigger mechanism, which would in effect allow domestic prices to continue to be considerably more stable than those on international markets. Exporters feared that this would prove to offer little improvement over the present variable levy system of the EC, and indeed encourage other importers to introduce such levies.[55] The US proposed a 'snapback' arrangement allowing a temporary increase in the height of the tariff, triggered either by world price movements or by import surges. In the end both types of safeguard mechanisms were agreed, as alternatives, in the case where non-tariff import measures have undergone tariffication.

Exporters asked for assurances that tariffication would not reduce access in the short run as a result of tariffs being introduced at high levels. The US argued for minimum access agreements, based on historical trade flows or on an agreed percentage of consumption, which would expand over time. The same quantitative import obligation would substitute for tariffication in cases where conversion to tariffs was not possible. Such access guarantees posed problems for countries that had no mechanism to enforce such minimum import levels, and the EC in particular was reluctant to agree to any expanding tariff quota for imports.

The tepid acceptance of tariffication by the EC left the opposition to this rule change to shift to other countries. One aspect of this change caused some controversy within the ranks of the exporters. That issue was the fate of the clause in Article XI:2(c) of the GATT, which allows quantitative restrictions on imports if in support of domestic supply restraints. The US and most of the Cairns Group saw the clause as a loophole to limit imports, with no effective way of restricting its application to cases where domestic supply control is effective. Canada, with an extensive network of marketing boards which attempts to manage supply, argued for a continuation of the exception albeit with 'clarifications', in particular on the issue of product coverage.[56] Disagreement on this issue significantly weakened the cohesion of the Cairns Group in the later stages of the negotiations. Japan also showed an interest in preserving Article XI:2(c), arguing at one time for its expansion to cover 'food security' issues, and the EC at one point even suggested that supply restraint through policies other than quantitative controls could be allowed to qualify for quantitative import controls. In the end, rendering moot the contentious clause in Article XI (though it technically still exists) was one of the main benefits of the agreement on tariffication.

Even more strenuous opposition to tariffication came from the rice producing countries, particularly Japan and Korea. The problem which these countries faced was in part a function of their agricultural price policies and in part a product of domestic politics. Japan and Korea, along with a few other countries in East Asia, maintain very high rice prices as a matter of rural policy, implemented by state trading entities which control both domestic and trade markets. The political strength of the farm lobbies was effectively mobilized to preserve this structure. Tariffication appeared to threaten the status quo, although the high level of tariffs which would replace the quantitative import controls would still have given substantial protection to the farm sector. As a way of squaring the circle, Japan and Korea were allowed to delay the implementation of tariffication for some time, agreeing to a larger but still limited market opening in its place.

One group which found no problem with tariffication, and indeed anticipated the rule change by unilateral action, was that of those Latin American countries that had restructured their trade policies as a part of general economic reforms. Led by Chile, more than a dozen countries adopted neo-liberal economic policies in the 1980s. Among the elements of the new economic package adopted by these countries was the conversion of non-tariff import barriers to tariffs. Other parts of the overall reform package made this change attractive. Privatization of trade and the removal of state trading were important aspects of the new policy. In a regime of state trading, import licences and quantitative restrictions through the rationing of such licences are the obvious form of import control. With private trade, tariffs have significant advantages in contrast to import licences. Thus a wave of tariffication, with new tariffs at low levels as required by the liberal economic philosophy, swept across the South American continent. The Uruguay Round process helped to lock this process in against the possibility of recidivism by successor regimes. That the Uruguay Round was considering the same move for improving the conditions of world trade was by way of a bonus.

The benefits of tariffication for the trade system were clear to all parties, even if individual countries had political difficulties with its implementation.[57] The tariff-only rule in the GATT was recognized as good for transparency and negotiability, but also as beneficial for efficiency, world market stability and an equitable sharing of adjustment needs. The possibility of extending the rule to agriculture, in essence by avoiding those parts of Article XI that allowed non-tariff measures to persist in agriculture and by reducing the scope for 'grey-area measures' such as variable levies, had been discussed in the GATT Committee on Trade and Agriculture set up in 1982. The 1985 US Farm Bill had identified the concept as a long-term aim of the US. But few would have thought it possible within the time frame of the Uruguay Round.

As compared to the interest generated in the use of an AMS, rather little consideration was given, by academics or officials, to the modalities of tariffication. The 'guidelines' for the process of tariffication were contained in a document which, though agreed by negotiators during the talks, had no place in the Final Act.[58] Instead, the process of tariffication was left to individual countries to draw up schedules, with only those of the major actors being subject to any detailed scrutiny and negotiation. Most other countries were able to submit tariff schedules that gave them adequate room to run current domestic policies and imposed little in the way of discipline in the near future.[59] While tariffication remains an important rule change with potential for improving market access and constraining domestic policies, further action will be needed to ensure that this potential is realized.

Control of Export Subsidies

The Uruguay Round negotiations struggled mightily with the issue of export subsidies. The Cairns Group countries had made this a major plank in their own proposals, and their cohesion rested in no small part on their concern about the export subsidy war which rumbled on between the US and the EC in the second half of the 1980s. The outcome was therefore a little disappointing for these countries. In place of a ban, a constraint on the total expenditure on export subsidies and on the quantities of exports eligible for such subsidies was negotiated. In partial response to the demands of the self-proclaimed 'non-subsidizing exporters', countries agreed to prohibit the granting of export subsidies to products not receiving such subsidies in the base period. As a result, one might expect the problem to improve over time, as the constraints begin to bite. However, the potential for considerable market disruption by countries with surpluses still exists, and the situation in agricultural goods still differs from that in manufactured trade.

Categorizing Domestic Subsidies

The direct negotiation of import access and of limits on export subsidies exposed a third category of policies which can distort trade. Domestic subsidies, such as deficiency payments, transport subsidies, input subsidies, marketing grants and capital subsidies, all have an impact on trade to the extent that they encourage production. Though the GATT articles do not make much of a distinction between export subsidies and domestic subsidies with similar effects, in practice the issue of domestic farm subsidies had not been successfully addressed in the GATT up to the time of the Uruguay Round.

The US and the Cairns Group were in broad agreement on the way in which to deal with domestic subsidies: they should be categorized into those that are acceptable ('green light') and those that are objectionable ('amber light'), with the latter subject to reduction over time.[60] The EC was concerned during the negotiations that too many of the policies of its competitors would end up as 'green', such as the US deficiency payments programme for grain. By contrast the EC used very few deficiency payments or direct subsidies, the main exceptions being subsidies to oil-seed, olive oil, and durum wheat producers. Accordingly, the EC suggested at one point that reduction commitments cover all domestic subsidies. The situation changed drastically with the adoption of the main parts of the MacSharry Plan for CAP Reform in May 1992. CAP reform introduced headage payments as compensation for the fall in support prices. In the end, the EC was content to allow its own compensation payments to be allocated, along with the deficiency payments of the US, to a 'blue box' of dubiously green policies which were both better than their predecessors and politically difficult to reduce.

The importance of the green box is twofold. It provides a set of internationally approved guidelines for policy changes, and it allows such policies to be excluded from the calculation of total support. It may be that the guidelines are too closely tied to conditions in the most affluent countries, with their emphasis on the payment of decoupled payments to farmers as an alternative to output-increasing price supports. Not all countries are struggling with agricultural problems characterized by the overcommitment of resources as a result of high support prices. Those countries that have problems of underinvestment in rural sectors as a result of infrastructural weakness, lack of market signals or imperfectly operating factor markets are less likely to be impressed with the new rules. However, developing countries have looser criteria for policies to be included in the green box, and in any case do not have to reduce the level of support to the same degree.

The Scorecard

It is tempting to ask who won and who lost in the long-drawn-out negotiations on agriculture. Presumably all trading countries won the benefits of a continued, rule-based multilateral system in the Uruguay Round. That system should be stronger not just because of the more stable institutional arrangements and the improved procedures for settling disputes but also because of the broader sector coverage. The negotiations on agriculture were a key part of the Uruguay Round from start to finish. Reaching an agreement on agriculture took time, but it was finally accomplished. Unlike the two previous rounds, the temptation to cut a quick deal on agriculture to allow the rest of the Round to be completed was resisted. Although the US toned down its demands for a dramatic liberalization of agricultural trade it remained adamant about the full inclusion of agriculture. The EC seemed to believe until the later stages of the Round that the US would finally settle for something less. In the end, the EC blinked, and the reform of agricultural trade policy rules was accomplished.

For this outcome the Cairns Group was in part responsible. It was the motley group of 'non-subsidizing' agricultural exporters which kept the US and the EC with their feet to the fire. The leverage was not so much their agricultural weight, though they accounted for a significant part of world agricultural trade. They in effect provided the glue that prevented the Round from degenerating into a North–South confrontation. They were able to hold out for the agricultural changes that were a *quid pro quo* for the inclusion of services. With the Latin American countries showing stiff resolve in the face of attempts to placate them with special market access deals, the brashness of Australian and New Zealand representatives in keeping their simple message before the negotiators, and the increasing political weight of the Asian members, the Cairns Group was able to deliver. They did not represent all developing countries, but they effectively provided a third voice to join with those of the US and the EC, and in the process empowered other countries in the negotiations. However, the Cairns Group was disappointed in its attempt to end export subsidies in agriculture. This alone gives it a reason to continue in existence and keep up the pressure on the US and the EC.

The success of the US in the Round in agriculture is less easy to define. The CAP was a major target, as had been the case for thirty years. The CAP has been pushed, by internal and external pressures, down a path towards domestic sustainability and international responsibility. If CAP reform is included as an outcome of the Round then one has to say that the outcome was of real and lasting value to the US and other agricultural traders. If one takes the view that this reform would have happened in any case, the Round has less to show for the efforts of the negotiators.

These changes in national farm policy illustrate the problem of attributing simple causes to complex events. The Uruguay Round negotiations were only one of a number of parallel processes that were going on in the agricultural policy arena at the time. Radical reforms in farm policy were being undertaken in country after country in the late 1980s, as a result of paradigmatic changes in economic policy. Regional trade arrangements were also chipping away at isolated agricultural markets in many parts of the world. It may be enough to say that the Uruguay Round gave a sharp stimulus to countries to change their policies, helped to determine the direction of this reform, and locked in those changes for the next few years. Domestic and trade policy forces converged in a way that made each more effective. US negotiators, however, must have been disappointed by the modest degree of trade liberalization which was accomplished. Some increased market access was achieved, but in their eyes possibly not enough to warrant seven years of intense diplomatic effort. The US crusade for open markets is unlikely to be at an end.

The EC escaped without having to dismantle the CAP or promise eventual free trade. Moreover, the reform programme in the Uruguay Round Agreement finally legitimizes the CAP in the GATT. The CAP is however markedly different now from the policy of 1986, and is unlikely ever to return to exclusive reliance on high market prices for farm income support. The EC clearly 'lost' in its effort to raise oil-seed and cereal substitute tariffs, as an act of rebalancing, but equally could be said to have escaped with the huge domestic oil-seed sector protected from further challenge under the GATT. Tariffication will have relatively little immediate impact, at least in cereals: the last-minute agreement to put a limit on the tariff-inclusive price of imports at the level of the existing threshold price essentially obliges the EC to keep a variable levy system. The EC seemed to have little in the way of positive objectives for changes in other countries, preferring to defend its own policies, but the removal of the US waiver could be called a successful outcome for the EC. The 'peace clause' seems to give the CAP a breathing space for the next few years, and to remove some of the external pressures on commodity programmes. However, the fundamental issues have not been resolved. The EC still has considerable overcapacity, and is not yet competitive in the temperate zone commodities. It is difficult to see how future trade problems can be avoided unless these internal problems are fixed.

8 The Uruguay Round Agreement on Agriculture

As befits a negotiation that lasted over seven years, the outcome of the Uruguay Round (UR) is contained in a document of nearly 500 pages.[1] The Agreement on Agriculture is only a small part of this text, with just 26 pages, and the Agreement on the Application of Sanitary and Phytosanitary Measures has a further 15 pages. However, an important part of the UR Agreement on Agriculture comes in the form of the quantitative commitments contained in the country Schedules which form part of the overall agreement reached in the UR. Taken together for all 117 countries which signed the UR agreement, the agricultural parts of these Schedules amount to around 20 000 pages.

With this in mind it is clear that the present chapter can only be an introduction to the agricultural agreement reached in the UR. We intend in this chapter to provide an overview of the economic provisions which were agreed and of some policy implications they may have. We shall start with a brief description of the overall structure of the Agreement on Agriculture, and then move to the three major areas of disciplines agreed: market access, export competition, and domestic support. In each of these areas, we shall outline the provisions agreed in some detail, comment on the nature of country commitments in the Schedules, and discuss issues of implementation. The Agreement on the Application of Sanitary and Phytosanitary Measures is discussed in a separate section. Even though the treatment of agriculture in the GATT has now been made much more similar to that of industry, some notable differences remain, and new ones have been created. These will be analysed in the final section of this chapter.

1 STRUCTURE OF THE AGREEMENT

The legal texts agreed in the UR have a complex structure. The overall agreement reached in the negotiations is confirmed in the short Final Act, giving effect to the various results of the UR. Among those results, the central element is the Marrakesh Agreement Establishing

The World Trade Organization. This Marrakesh Agreement has a number of annexes. Annex 1A contains the Multilateral Agreements on Trade in Goods. These Multilateral Agreements are the 'GATT 1994' and 12 new agreements negotiated in the UR, including the Agreement on Agriculture and the Agreement on the Application of Sanitary and Phytosanitary Measures. An important adjunct to GATT 1994 is the set of country Schedules in which all WTO members specify their tariff bindings, non-tariff concessions, and their commitments regarding agricultural subsidies.

There is now a distinction between the 'GATT 1947' and the 'GATT 1994'. The GATT 1994 is not a completely rewritten agreement. All the articles of the 'old' General Agreement (now called GATT 1947) are still alive. Interpretation of some of them has been amended through a number of understandings which now form part of GATT 1994. Moreover, the new agreements contained in Annex 1A of the WTO Agreement have been added. In some cases this means that there are apparent contradictions between the text of the General Agreement (still being the 'old' text) and the more recently agreed provisions. For example, Article XI:2(c) of the 'old' GATT still exists, saying that quantitative restrictions to imports can be used in agriculture if certain requirements are met. On the other hand, the UR Agreement on Agriculture bans the use of non-tariff measures, except in a very few country-specific cases. Wherever such conflicts between the 'old' GATT and the new agreements exist, the new agreements take precedence. This is explicitly stated in an interpretative note to Annex 1A of the WTO Agreement. In other words, rather than rewriting the text of the General Agreement, UR negotiators have agreed new provisions which automatically substitute for the 'old' rules wherever necessary.

In agriculture, this effectively means that anything that was a special provision for this sector in the 'old' GATT has now been superseded by the respective provisions in the new Agreement on Agriculture. Moreover, some new special rules for agriculture have been introduced, which are also contained in that Agreement. In addition, there is now agreement to reduce tariffs and subsidies by certain percentages. However, these percentages cannot be found in the Agreement on Agriculture. The reason is that negotiators were wise enough to relieve the process of implementing the Agreement[2] of potential disputes about determining the starting point for reductions and the resulting commitments for future years. To this end, they agreed during the negotiations what the reduction percentages should be, how the bases should be established from which the reductions have to be made, and how the calculations should be done. All this was laid out in a working document, called the 'Modalities' paper, which specified both the reduction percentages and the calculation methods, as well as other technical details regarding the commitments to be undertaken.[3] On the basis of this working document, it was then left to each country to calculate its suggested commitments on such matters as tariff rates and export subsidy constraints. The results of these calculations were inserted, in an agreed format, in offers tabled in Geneva, and the intention was that countries would check each other's offers during the process of verification in the last few weeks of the Round (see Chapter 7). Once offers were agreed, the respective figures became part of each country's Schedule under the Agreement Establishing the WTO, and thereby an element of international law. As a consequence of this approach, the Agreement on Agriculture itself contains almost no numbers. However, its text explicitly refers to the country Schedules. Though technically not quite correct, in the following pages we shall often use the term 'Agreement on Agriculture' as referring to both the text of the Agreement and the agricultural components of all country Schedules.

The structure of the various provisions in the Agreement on Agriculture is best understood by considering the three areas on which negotiations focused, namely import access, export competition, and domestic support. In each of these three areas, new rules were defined and rates of reduction were agreed. This pattern is reflected in Table 8.1. The details of all these provisions will be discussed in the subsequent sections.[4] As far as market access and export competition are concerned, both the new rules and the reduction requirements come in price-related and quantity-related forms. Commitments regarding export competition and domestic support are jointly covered by a 'peace clause'.

An institutional innovation is the new Committee on Agriculture which has been established under the Agreement. One of the main tasks of the Committee will be to review the implementation of the Agreement. In this context, new notification requirements have been established. Matters relevant to the implementation of the Agreement can be discussed in the Committee. However, formal consultation and disputes among Members will be governed by the new obligations of the Understanding on Rules and Procedures Governing the Settlement of Disputes.

The implementation period under the Agreement is the period from 1995 to 2000. An important element of the Agreement is its provision for the 'Continuation of the Reform Process'. Countries have agreed to continue the process of reductions in support and protection. For

Table 8.1   Structure of the Agreement on Agriculture[a]

Type of rule	Market Access (Base: 1986–8)	Export Competition (Base: 1986–90)	Domestic Support (Base: 1986–8)
Price	Tariffication of NTM	Reduction of outlays on export subsidies by 36% (product specific)	
	Reduction of new tariffs by 36% on average (minimum of 15%)		Reduction of total AMS by 20%, except for 'green box' measures
Quantity	Minimum access commitments: 3% of domestic consumption, growing to 5%	Reduction of subsidized exports by 21 %	
	Current access maintained		
Other	Safeguard provisions	Peace clause	

[a] Reduction rates in this table are those for industrialized countries; for developing countries, reduction rates are two-thirds of those for developed countries, and the implementation period is ten years. Least-developed countries are exempt from reduction commitments.

this purpose, negotiations will be initiated one year before the end of the implementation period (that is in 1999). In this 'mini-round', experience gained with implementing the Agreement will be taken into account, and further commitments will be considered.

As an incentive for countries to accept the new disciplines and commitments on domestic support and export subsidies, it was agreed that subsidies that conform to the new rules are sheltered from international challenge under the GATT. The Due Restraint provisions of the Agreement, generally referred to as the 'peace clause', state that 'green box' policies are non-actionable for purposes of countervailing duties and other GATT challenges; that all domestic support which conforms with commitments, including payments under production-limiting programmes (that is US deficiency payments and EC compensation payments) is countervailable (if causing injury, and 'due restraint' shall be shown in initiating countervailing duty investigations), but exempt from other

GATT challenges as long as the AMS for the commodity concerned does not exceed that decided in 1992; and that export subsidies within the constraints of the Agreement are countervailable (as is domestic support outside the green box), but exempt from other GATT challenges. It is hoped that this peace clause will limit the number of disputes in the area of agricultural trade. The Due Restraint provisions remain in force three years beyond the implementation period of the Agreement, that is until the year 2003.

2 MARKET ACCESS

Before the UR, the use of non-tariff measures was widespread in agricultural trade. Moreover, in many countries the particular types of NTM used had strategic importance for the functioning of their domestic market regimes. Against this background, the UR agreement to improve market access in agriculture, by converting all NTM into bound tariffs and by binding all previously unbound tariffs, must be counted a genuine breakthrough. However, the rule changes agreed for market access in agriculture will have a less than revolutionary practical impact on agricultural trade, at least during the first years of the implementation period of the Agreement. Moreover, some of the more immediate effects of the changes agreed for market access may not be exactly in line with what a more liberal trading regime would require. In particular, implementation of the minimum access provisions can cause undesirable economic effects. All this results from the way in which Agreement provisions on market access have been translated into Schedule commitments and from problems which may occur during the implementation of the Agreement. However, none of these problems eliminates the longer-run benefits which will flow from the new rules on market access in agriculture.

Agreement Provisions

The basic provisions on market access in the Agreement are rather short. Their central element is one sentence in Article 4:2 which provides that 'Members shall not maintain, resort to, or revert to any measures of the kind which have been required to be converted into ordinary customs duties'.[5] In other words, tariffication of all NTM is required. To eliminate the grey area for NTM which was so widespread before the UR, a footnote to this provision enumerates the most important non-tariff border measures that now have to be converted into tariffs. It says that 'these measures include quantitative import restrictions, variable import levies, minimum import prices, discretionary import licensing, non-tariff measures maintained through state trading enterprises, voluntary export restraints and similar border measures other than ordinary customs duties'. By listing this now prohibited arsenal of traditional agricultural trade barriers, and by allowing only 'ordinary customs duties', the Agreement is much more specific than Article XI:1 of the GATT[6] which bans 'prohibitions or restrictions other than duties, taxes or other charges'. It is also important to note that the footnote indicates that all border measures need to be converted into tariffs, irrespective of 'whether or not the measures are maintained under country-specific derogations'.[7] Thus no grandfather clause, protocol of accession or waiver can any longer be used for justifying a non-tariff barrier in agriculture. As one of the results, the 1955 US waiver (see Chapter 2) is now a thing of the past.

The huge importance which this short sentence and the footnote to it have is that (nearly) all NTM have now disappeared in agricultural trade. All countries had to engage in tariffication because they may not continue to apply ('not maintain') traditional NTM, they may not introduce new ('not resort to') NTM, and they also may not at some stage fall back from tariffs to ('not revert to') their past NTM. In addition, paragraph 7 of the Modalities required that 'all customs duties, including those resulting from tariffication shall be bound'. It is difficult to think of a more stringent expression of the 'bound tariffs only' principle.

There are only two exceptions to this 'bound tariffs only' obligation. First, on the insistence of Japan and Korea, a 'rice clause' was added to the Agreement, as Annex 5 ('Special Treatment under Article 4:2'), allowing members to postpone tariffication for some time under certain conditions. These conditions are phrased in general terms, but such that they cover the rice policies of these two countries. The rice clause comes in two forms, for developed countries (Japan) and developing countries (Korea). For developed countries it allows tariffication to be postponed until at least the end of the implementation period for the Agreement (1995 to 2000), and possibly beyond if this can be agreed in negotiations and if the country concerned makes other concessions. As a *quid pro quo*, the importing country has to open up, for the product concerned, a tariff-rate quota that is larger than would be required under the general provisions regarding minimum access. The quota has to start at 4 per cent of base period domestic consumption and grow to 8 per cent by the end of the implementation period. For developing countries, tariffication can be postponed for ten years (and beyond if successfully negotiated), and minimum access starts at 1 per cent and grows to 4 per cent of base period domestic consumption. When special treatment under the rice clause expires, tariffication has to take place, essentially in the way in which it applied in all other cases.[8] Where countries wanted to make use of the rice clause, they had to declare this in their Schedule. In addition to Japan and Korea for rice, this was also done by the Philippines for rice (developing country clause) and by Israel for sheep and goat meat, and for some cheeses and milk powder (developing country clause). Since no other country invoked the Special Treatment clause, its use will be limited to these few cases. In all other cases, tariffication actually took place at the beginning of the implementation period.

The second exception from the 'bound tariffs only' principle is the possibility of imposing additional duties in cases of import surges or world price drops. This possibility, regulated in Article 5 ('Special Safeguard Provisions'), is limited to products where tariffication has taken place and where the countries concerned have reserved the right to invoke this clause by explicitly designating these products in their Schedules. Countries have nearly universally designated products under tariffication as eligible for the special safeguards. The Special Safeguard Provisions (SSG) come in a quantity-triggered and a price-triggered form, but only one of them can be invoked at any time. The quantity-triggered additional duty can reach a maximum of one third of the ordinary bound duty in effect for the product concerned. It can be levied in any year, until the end of the year, in which the volume of imports exceeds a given trigger quantity. The trigger quantity is calculated, through a somewhat cumbersome formula, such that the extra duty can be levied if imports grow by more than a given rate over their past three year average. This rate of growth is higher the lower the share of imports in domestic consumption is, and it is higher the more domestic consumption grows.

The price-triggered SSG relates the size of the additional duty to the extent to which the c.i.f. import price drops below a given trigger price, again through a formula which could come straight from a university seminar on the economics of price stabilization. The trigger price is the world price in the base period. It 'shall, in general, be the average c.i.f. unit value of the product concerned' in the 1986–8 period, in the domestic currency of the importing country. The additional duty increases as the c.i.f. price drops below that trigger price, and it becomes an increasing percentage of that price drop as the c.i.f.

price goes down. The formula for calculating the additional duty is such that it compensates for some of the effect which a world price drop might otherwise have on the price at which imports enter the country. In that sense there is some similarity between the price-triggered additional duty and a variable levy, though the additional duty under the formula adopted in the Agreement is less variable than traditional variable levies (which compensate for all of the variability in the world price). However, it is important (and disturbing) to note that the price-triggered additional duty is shipment-specific, that is it depends on the billed price of each individual consignment. In this regard it is different from traditional variable levies which used to be charged at a given rate on all imports coming in during a given period.

There are, in addition, other important elements of the agreed provisions regarding market access which are not reflected in the text of the Agreement. In particular, criteria were agreed under which countries were expected to calculate the tariff equivalent of their former NTM. These tariff equivalents were supposed to become the bound tariffs for the first year of the implementation period (1995). It was also agreed at what level previously unbound tariffs were to be bound. Moreover, it was agreed that all tariffs, old and new, were to be reduced by 36 per cent as a simple (that is unweighted) average during the implementation period, in equal annual steps. No single tariff, though, was to be reduced by less than 15 per cent.[9] All these important provisions were included in the Modalities, not the Agreement as such, and in the country commitments resulting from them as incorporated in the Schedules.

Another set of important provisions not explicitly included in the text of the Agreement relates to 'minimum access' and 'current access'. The conditions governing this part of access commitments have initially been agreed in a form which is reported in the Modalities document, and the respective country commitments are embedded in the Schedules. The only indirect mention which the text of the Agreement makes of this part of the market access provisions is reference to 'other market access commitments as specified' in the Schedules (Article 4:1). Both minimum access and current access commitments come in the form of tariff-rate quotas, allowing imports of specified quantities at tariffs lower than the 'normal' bound tariff rates. Both quota volumes and within-quota tariffs are specified in the Schedules, and in a legal sense they are equally as binding as the bound tariffs.

Minimum access opportunities were required to be established 'where there are no significant imports' (Modalities, paragraph 5). The size of

the respective tariff-rate quotas was agreed to equal 3 per cent of domestic consumption in the 1986–8 base period for the first year of the
implementation period (1995), growing to 5 per cent of that base period consumption in the last year of the implementation period (2000).
Within-quota tariffs were supposed to be 'low or minimal' (Modalities,
Annex 3, paragraph 14), but no general formula was agreed as to what
that meant in quantitative terms. Tariff-rate quotas resulting from minimum access commitments are to be allocated on a most-favoured nation
(MFN) basis.

Current access opportunities were supposed to be made where the
process of tariffication would otherwise have resulted in market access
conditions less favourable than those existing in the base period. In
such cases, tariff-rate quotas were to be established at quantities imported during the 1986–8 base period.[10] In many cases, countries argued that the tariffs resulting from tariffication were lower than the
levels of protection in the base period and that therefore they did not
have to make current access commitments, and they got away with
that argument. As a result, current access commitments in the Schedules relate mostly to cases where special conditions had been granted
for certain imports. For example, where countries established import
quotas and allocated them to specific exporting countries, allowing the
exporters to gain (part of) the quota rent, simple tariffication would
have worsened access conditions for these exporters' because it would
have effectively transformed exporters' rents into tariff revenue for the
importing country. A case in point are US import quotas for cheese
which used to be allocated to specific exporting countries. Similarly,
where voluntary export restraints existed, straightforward tariffication
would also have eliminated the exporters' rents. Hence, voluntary export restraints, such as for example the arrangement regarding manioc
between Thailand and the EC, were also to be converted into tariff-
rate quotas, at pre-UR tariffs. The same is true for preferential trading
arrangements existing before the UR. An example is the preferences
for sugar imports the EC had granted to certain developing countries
under the Lomé convention. Other cases covered under current access
provisions include those where an importing country had committed
itself to allowing imports of a given quantity at conditions more favourable than normal imports. An example is the EC commitment to
import 2 million metric tons of maize into Spain. This commitment
had been negotiated bilaterally with the US in the context of Spanish
accession to the EC, but the EC import commitment had then been
multilateralized. In a case like that, the resulting tariff-rate quota is

not allocated to given exporting countries but has to be provided on an MFN basis.

Current access and minimum access are closely interrelated: where current access was less than the required minimum access opportunity, access was to be expanded. No distinction is noticeable between current access and minimum access in most of the Schedules. Both provisions have resulted in tariff-rate quotas, and there was no requirement to label these quotas according to their historical origin.[11]

Schedule Commitments

As far as market access for agricultural products is concerned, the Schedules agreed during the UR consist of two parts: Section IA lists the tariffs, and Section IB specifies the tariff-rate quotas. In the tariffs section, countries list the 'base rate of duty', applicable in the first year of the implementation period (1995), and the lower 'bound rate of duty', valid for the final year of the implementation period (2000). During the implementation period, duties are reduced in equal annual steps in order to reach the bound rate in the last year. Where a country had a bound tariff in the past, the Schedule shows this rate as the 'base rate', and the new 'bound rate' is the old rate minus the reduction agreed. Where a country had not bound a tariff in the past, but had applied an ordinary customs duty, the 'base rate' was supposed to be the rate applied on 1 September 1986 (Modalities, paragraph 3). Developing countries were allowed to offer 'ceiling bindings' for their earlier unbound customs duties, in effect meaning that they could bind tariffs at levels higher than those they applied in the past.

Where tariffication had to occur, Annex 3 of the Modalities described the approach to be followed in establishing the 'base rate of duty'. What was supposed to happen was essentially the calculation of the tariff equivalents of the NTM used in the base period, very much along the lines that a textbook on international trade might describe.[12] For the base period 1986–90, countries were supposed to calculate the tariff equivalent as the difference between the internal and the external price for the product concerned. The internal price, it was agreed, 'shall generally be a representative wholesale price ruling in the domestic market or an estimate of that price where adequate data is not available'. 'External prices shall be, in general, actual average c.i.f. unit values for the importing country'. Where actual c.i.f. unit values were 'not available or appropriate', import prices of a near country or export prices of major exporters, plus insurance, freight and other cost, could be used. Where necessary, adjustments could be made for quality differences. In the Schedules, the duties resulting from tariffication can no longer be distinguished from those that were already bound before the UR.[13]

It is important to note that the tariff equivalents were not calculated by a neutral institution but by the importing countries concerned. It did not come as a surprise that importing countries exhibited a tendency to maintain protection by trying to get away with as high initial tariffs as possible. Hence the internal and external prices used for calculating tariff equivalents for the base period tended to be chosen with this in mind. What may be more surprising at first glance is the fact that the verification process, following the submission of the draft schedules, was not used more by the exporting countries to correct these tendencies. Very few changes to the tariff equivalents tabled were made during verification.[14] Presumably the principle of tariffication and the binding of tariffs was so important to the exporting countries, and so difficult to accept by the importing countries, that it was felt risky to place too much pressure on importing countries to bind their new tariffs at low rates. In this sense it can be said that the rule change achieved, that is tariffication, was bought at the expense of more immediate liberalization, though it is also not clear whether more reductions in trade barriers could really have been achieved had the principle of tariffication not been pushed. In the end, tariffication was not pursued as a mechanical process of calculating tariff equivalents in a purely objective sense, but as a process of offers and (a few) requests, more or less loosely guided by the rules laid down in the Modalities.

An interesting example of the *ad hoc* nature of the tariff setting was the process of establishing tariffs for cereals in the EC. The EC submitted to the GATT secretariat its first offer for tariffication in response to the Dunkel Draft. For cereals, the EC Commission used as external prices f.o.b. prices in exporting countries, adjusted for transport cost and quality differences. As internal prices, the EC Commission used the EC intervention price, plus 10 per cent, plus all monthly increments until the end of the crop year. These prices were not completely unreasonable, but also not exactly what an independent analyst might have chosen. However, as part of the Blair House Accord, the US negotiators accepted that approach, and the EC Commission proudly announced their 'victory' at home (EC Commission, 1992). Later, the US Administration felt that the cereals tariffs resulting from that approach would not really improve access to the EC market, and might actually worsen it given the large reduction in the level of price support that the EC was going to implement as part of the MacSharry CAP reform. In the last-minute negotiations, on 6 December 1993, the US therefore persuaded the EC to accept an additional commitment not to use the full tariff resulting from tariffication. More specifically, in its Schedule the EC has undertaken, for most cereals, 'to apply a duty at a level and in a manner so that the duty-paid import price for such cereals will not be greater than [155 per cent of] the effective intervention price'.[15] Similar negotiations on the tariff rates resulting from tariffication have been conducted in other important cases, though it appears there is no other case where a country has undertaken not to use the full tariff bound in its Schedule.[16]

Some observers have called the outcome of this process 'dirty tariffication'. How 'dirty' the tariffs resulting from tariffication really are, in the sense of being higher than the actual equivalent of the base period NTM, is impossible to say in the absence of a detailed empirical analysis of market data. In any case, many tariff rates which have resulted from tariffication are indeed very high in absolute terms. Initial tariff rates (the 'base rates of duty' for 1995) are shown in Table 8.2 for some selected countries and products, expressed as *ad valorem* percentages.[17] Tariff rates of several hundred per cent are not uncommon. Dirty or not, these tariff rates certainly reflect the very high levels of protection which governments provided to their farmers through the NTM which were previously used in agriculture and their reluctance to reduce this protection at the time of tariffication.

The real test of the significance of the tariffs resulting from tariffication will be in the trade flows they allow or hinder. Where NTM were essentially prohibitive in the past, the initial tariff rates established are likely to be prohibitive as well. In particular, where countries relied on quantitative import restrictions before the UR, their initial tariffs tend to be so high that it is unlikely that much will be imported beyond minimum and current access quotas. For example, tariffication of US import quotas maintained under the Section 22 waiver (see Chapter 2) was expected to have virtually no effect on US imports of cotton and dairy products, beyond the minimum access commitments (Office of Economics, 1994). Equally, the high tariffs which have replaced Canada's import quotas for dairy products are likely to be prohibitive (Josling et al., 1994, p. 55). Where tariffication has taken place in this way, the actual trade situation will not change immediately as a result of the elimination of NTM. Indeed, there are cases where the 'base rates of duty' are higher than tariffs or levies actually applied immediately before the implementation period (1995), and in these cases

Table 8.2   Tariffs Resulting from Tariffication under the UR Agreement on Agriculture (*Ad Valorem* Equivalents, Selected Countries and Products)

	Common Wheat		White Sugar		Beef Carcases, Fresh or Chilled		Butter	
	Initial Tariff (%)	Reduc- tion (%)	Initial Tariff (%)	Reduc- tion (%)	Initial Tariff (%)	Reduc- tion (%)	Initial Tariff (%)	Reduc- tion (%)
Australia	0	–	31.7	50	0	–	4.6	78
Canada	90	15	10.7	15	37.9	30	351.4	15
EC	142.3	36	207.1	20	96.9	36	235.3	36
Hungary	50	36	80	15	112	36	159	36
Korea	10	82	94.6	10	44.5	10	99	10
Japan	422.9	15	326.7	15	93	46	97.7	15
New Zealand	0	–	0	–	0	–	10	36
Poland	143.2	36	120 .	20	162	36	160	36
Switzerland	477.6	15	159.9	15	139.7	15	862.2	15
USA	6	55	134.7	15	31.1	15	116.7	15

Sources: Deutsche Bundesbank (1994), EUROSTAT (1994), GATT (1994), IMF (1994), US Department of Agriculture (1995).

protection increased as a result of the Agreement.[18] However, with future reductions of tariff rates, during the implementation period of the UR agreement and possibly beyond, the large amount of 'water' which is in many of the newly bound tariffs will evaporate, and may eventually disappear altogether, such that trade can begin to flow.

The extent to which this is the case depends on the size of tariff reductions countries have agreed to make by the end of the implementation period. The agreement to reduce tariffs by an unweighted average of 36 per cent, with the only constraint being that reduction rates needed to be at least 15 per cent, left much freedom as to which products to treat in which way. In particular, since the reduction requirement was specified as a given percentage rather than a given number of percentage points, countries could find an easy way out by reducing low tariffs more than high tariffs. For instance, it is possible to reduce, say, tariffs on three items with initial tariffs of several hundred per cent by only 15 per cent each, and still meet the overall 36 per cent unweighted average reduction by eliminating (that is reducing by 100 per cent) a 4 per cent initial tariff on a fourth product. Where the required average reductions were achieved in such a way, the profile of protection looks more unbalanced after the UR than it did before, with potentially damaging implications for trade distortions.[19] For the

Agriculture in the GATT

Table 8.3   Average Tariff Reductions under the UR Agreement on
Agriculture (Selected Countries)

	EC	Japan	USA
1 Base period average tariff level (%)	26.2	52.3	11.3
2 Final period average tariff level (%)	17.7	40.2	7.9
3 Average of reduction rates (%)	37.7	36.8	38.8
4 Difference (1) − (2) as % of (1)	32.4	23.2	30.0
5 Coefficient of variation of base tariffs (%)	163.7	399.8	213.6
6 Coefficient of variation of final tariffs (%)	169.6	426.4	259.0

Source: Tangermann (1995).

countries and products selected, Table 8.2 also shows the rates of reduction included in the Schedules, confirming to some extent the hypothesis that low tariffs are scheduled to be reduced more than high
tariffs.

A more detailed picture of tariff reductions in three selected developed countries is provided in Table 8.3. All tariffs on agricultural and
food products in the countries covered are included in this analysis,
expressed as *ad valorem* equivalents.[20] Average tariff levels are calculated as unweighted averages. The unweighted average of reduction
rates (row 3) is close to the required 36 per cent in each of the three
countries. However, average tariff levels will decrease by less because
of a tendency to reduce low tariffs more than high tariffs (see row 4).
As a result, tariff dispersion (expressed as coefficient of variation of
all tariff rates) will increase in each of the three countries covered.

The minimum access and current access provisions have resulted in
a large number of tariff-rate quotas. For example, the EC Schedule has
51 tariff quotas, and the Canadian Schedule has 21 such quotas.[21] Countries have dealt in different ways with the need to establish tariff quotas under the minimum access provision. According to the Modalities,
minimum access opportunities were supposed to be established at the
four-digit level of the Harmonized System (HS), a relatively disaggregate
product level. However, this requirement was not always followed. For
example, in the case of meat the EC calculated the quota quantity
required under minimum access for the whole aggregate of 'meat',
comprising 18 product groups at the four-digit HS level. After having
calculated the total access opportunity in this way, the EC then allocated this quantity to individual types of meat at a more disaggregate
level, and the resulting tariff quotas for individual very specific products then became the legally binding commitments. As a result, the

EC could choose those types of meat where additional imports would hurt least, or where particular exporter interests could be benefited. These were sometimes qualities of meat not much produced in the EC, or meat categories for which the EC had already agreed preferential import conditions with Central European countries.[22]

The establishment of minimum access quotas was not a purely mechanical process, as might have been the case if left to an independent institution. In a number of cases it involved a request and offer negotiation, though within the framework of generally agreed rules. One indication of the bilateral negotiations which were behind some of the tariff-rate quotas is the fact that in some cases these quotas were larger than what a mechanical implementation of minimum access would have required, apparently because exporters had persuaded importers to provide better access.[23] Moreover, it appears that a significant number of bilateral side deals have been negotiated which are not reflected in the Schedules (and therefore not strictly enforceable under the GATT/WTO).[24]

In aggregate, increases in market access resulting from minimum access commitments agreed by all countries during the UR are not large relative to total world trade for most agricultural commodities. For some products, though, they are in the order of magnitude of 10 per cent and therefore have the potential of leading to a noticeable expansion of world trade in those commodities (see Table 8.4). However, whether minimum access commitments actually result in growing imports is not certain. Minimum access commitments are not guarantees to import, but are designed to establish opportunities for imports. In practice this means that within-quota tariffs under minimum access are not zero, but just reduced *vis-à-vis* the bound MFN tariffs. Countries have dealt very differently with the extent to which within-quota tariffs are reduced in their Schedules. For example, the EC has generally set its within-quota tariffs at 32 per cent of the normal tariffs. In Canada, the within-quota tariff for turkey meat is between 4 and 9 per cent of the normal tariff, while for butter it is 6 per cent. In Japan, within-quota tariffs for wheat and butter are zero. In some cases, within-quota tariffs stay constant over the implementation period (and therefore increase relative to above-quota tariffs), in others they are reduced.

Continued (or even new) state trading is not ruled out by tariffication under the Agreement. The Agreement only requires state trading enterprises to abandon practices that act as non-tariff barriers to trade. As a result, where countries intend to continue pursuing imports through state trading enterprises, they have typically bound the 'mark-ups' to

Table 8.4   Increases in Market Access Under Minimum Access
Commitments (Selected Products)

Product	Increases in Access Opportunities between 1986–8 and 2000	
	Million metric tons	*Proportion of 1992–3 World Trade[a] (%)*
Wheat	0.807	0.8
Coarse grains	1.757	1.9
Rice	1.076	7.4
Bovine meat	0.186	4.1
Pigmeat	0.133	7.7
Poultry meat	0.094	3.5
Eggs	0.252	1.7
Milk powder	0.147	6.9
Cheese	0.132	14.1
Sugar	0.292	0.9

[a] World trade excludes intra-EC trade.

Sources: EUROSTAT (1994), GATT Secretariat (1994), US Department of
Agriculture (1994), FAO (1995).

be collected by these agencies in their Schedules (e.g. Japan for a
number of products).

Implementation Issues

At first glance it may appear as if there will not be major issues regarding the implementation of the new rules and country commitments
in the area of market access. Tariffication has eliminated nearly all
NTM in agricultural trade and there should be few problems in implementing the tariffs which have replaced the former NTM. However,
this is not likely to be the case. A number of issues need to be considered, including the switch from variable levies to tariffs, the Special
Safeguard Provisions, and the administration of tariff-rate quotas.

The application of flat rate tariffs ('ordinary customs duties' in GATT
terminology) is required by the Agreement, and countries have committed themselves to tariffs in their Schedules. However, nothing requires that countries always fully use the tariff they have bound. Instead,
they can apply a lower tariff. If they do, some countries may wish to
adjust these below-commitment tariffs upward and downward when world
market prices fluctuate, so as to keep domestic market prices more
stable. The price band policies pursued in some Latin American coun-

tries are cases in point. In practice, such adjustable tariffs would be similar in effect to the variable levies which existed in the past. Bound maximum duties would have to be respected, but the average tariff charged over time would in such cases be below their admissible maxima. From the point of view of trade liberalization this would be beneficial. On the other hand, the continuation of measures similar to variable levies in practice would eliminate two of the hoped-for advantages of tariffication, its contribution to transparency in import barriers, and the advantage of fixed tariffs for the stability of world market prices. Issues may well arise as to the GATT-legality of duties which vary over time. Article 4:2, and the footnote to it, outlaw variable levies. An 'ordinary customs duty' that is adjusted from time to time could potentially still be considered to be an 'ordinary customs duty'. But if countries use a border regime of that type, they have in any case to observe general GATT rules which aim to give 'known and secure conditions of access', in particular GATT Articles I (MFN treatment), II (consistency with Schedules), and X (timely publication and uniform administration). In addition, adjustment over time of tariff rates must not be made in such a way such that it discriminates between different exporters.[25] Some lawyers have in the past argued that variable levies as used by the EC may have violated the requirement to publish tariffs (and other measures) 'promptly in such a manner as to enable governments and traders to become acquainted with them' (Davey, 1993, p. 6).

In this context it is interesting to consider the specific commitment the EC has undertaken in respect of its cereals tariffs, as mentioned above. The undertaking to apply tariffs, such that the duty-paid import price of cereals never exceeds the given threshold of 155 per cent of the EC intervention price, requires the EC to vary its cereals duties when world prices fluctuate, at least if the EC always wants to make full use of the maximum permissible protection. In this sense, the result of the last-minute bilateral negotiations between the US and the EC, in which this undertaking was agreed, essentially obliged the EC to maintain a system similar to its past variable levies, even though in prior negotiations these had always been characterized as a particularly detrimental form of trade policy. In one important respect, however, the EC commitment, taken literally, suggests that a shipment-specific duty should be collected rather than the old-style variable levy.[26] On a number of counts, a shipment-specific duty has economic effects which are considerably more disruptive than those of a variable levy (Tangermann and Josling, 1994). It is not clear whether negotiators were fully aware of these implications when they agreed on this particular undertaking by the EC.

Somewhat similar issues may arise out of the price-triggered Special Safeguard Provision (SSG). Even more explicitly than in the case of the EC undertaking in respect of cereal tariffs, this provision requires countries to set the additional duty on a shipment-specific basis. The implications will not be quite so severe, though, because the additional duty is less than 100 per cent of the difference between the trigger price and the price of the shipment concerned. This means that competition among exporters is not completely eliminated. However, there is still an incentive for the importing and the exporting companies to get together and agree on an invoice price which is above the trigger price. If this practice is adopted, the price-triggered SSG may not be very effective. It may also result in significant rents to private traders. Moreover, exporters who do not engage in such collusion with importing companies may feel discriminated against because they face the additional duty.

Another serious issue is the choice of trigger prices. In contrast to the external prices used for tariffication, which could be checked during the verification process because they were disclosed in the offers, there was no *ex ante* agreement on trigger prices to be used under the SSG. Determination of the trigger price is not expected to happen before the SSG is used for the first time. However, at that stage discussions may arise. The definition of the trigger price under the SSG reads very similar to that of the external price to be used for tariffication in the Modalities. Yet it may not always be possible to rely on the price used during tariffication, among other reasons because the SSG will be applied at a lower level of product aggregation than that at which tariffication occurred. The potential for conflict is illustrated by a list of trigger prices intended for use by the EC which was appended to the EC offer. Prices in this list are generally much higher than the external prices the EC used for tariffication. If used, these prices would allow the EC to avoid much of the intended implications of tariffication (Josling and Tangermann, 1995).

When it comes to the minimum and current access provisions, there are serious issues related to the allocation of licences under the respective tariff-rate quotas. These issues may assume significant importance in the implementation of the Agreement. After all, there is now a large number of tariff-rate quotas in agricultural trade, many of which did not exist before the UR. Administration of these quotas will not always be easy, and disagreement may well arise over the proper approaches to be adopted. The allocation of licences under tariff-rate quotas involves distribution issues at two levels. First, which exporting countries should be given access, and for what quantities? Second, how should licences be allocated to companies? The Agreement on Agriculture does not provide guidance on either of these questions. The Modalities paper suggested that minimum access opportunities 'shall be provided on an MFN basis' (Annex 3:14). However, the Modalities paper has lost legal standing after the verification process, and one now has to look for guidance in the general rules of the GATT.

As far as allocation to exporting countries is concerned, the relevant provisions are those in GATT Article XIII ('non-discriminatory administration of quantitative restrictions') which requires that countries 'shall aim at a distribution of trade . . . approaching as closely as possible the shares which the various contracting parties might be expected to obtain in the absence of [the quota]'. In doing so the country applying the quota 'may seek agreement with respect to the allocation of shares in the quota with all other contracting parties having a substantial interest', or where 'this method is not reasonably practicable . . . allot . . . shares based upon the proportions supplied . . . during a previous representative period'. In many cases, neither of these two options will be easily implemented. Negotiations with all contracting parties having a substantial interest will prove difficult. Reference to a previous period will not be possible where a market is opened for the first time (or significantly more than in the past) as may be the case in a number of minimum access commitments.

At the same time, allocation of licences to individual trading companies will not be straightforward. If significant rents are to be earned, companies will rush to apply for licences. Governments use different methods when it comes to solving this type of problem. One of the most frequently applied approaches is to allocate licences on the basis of shares of trade in a past reference period. As with the allotment of quotas to supplying countries, this method is unworkable where the market is opened up for the first time. Moreover, it does not promote economic efficiency.

Fortunately there is one approach that could be used to solve the problems at both levels. Licences under tariff-rate quotas could be auctioned by the government of the importing country. Any company, from any country, should be permitted to participate in these auction procedures. At the level of private trade, auctioning would eliminate the problem of allocating the right to earn rents. At the level of exporting countries, auctioning would tend to create 'a distribution of trade . . . approaching as closely as possible the shares which the various contracting parties might be expected to obtain in the absence of [quotas]'. There is, though, the issue of whether the collection of auction fees does not amount to the imposition of an extra charge on the imports concerned, which might not be GATT-legal. However, auctioning of quotas should be regarded as a method of internally distributing the licences, rather than collecting a charge on importation, and the revenue from the auction could be seen as capturing the rents from the application of a public policy for public use. From an economic perspective it is clear that the maximum fee a company would be prepared to pay in an auction is the rent which would otherwise accrue to it. In other words, collection of a fee through the auctioning of licences under tariff-rate quotas would not impose an extra economic burden on the exporting country, but simply skim off a rent which otherwise would have accrued to a private trader. Since access commitments have not been negotiated in the Agreement in order to benefit traders, there is no reason why importing countries should help them obtain these rents.

None of these implementation issues in the area of market access is serious enough to undermine the validity of the Agreement. Moreover, with flexibility on all sides, solutions can be found to all of these issues. However, this may require ongoing negotiations, which could constructively be held in the new Committee on Agriculture.

3 EXPORT COMPETITION

Curbing export subsidies in agriculture was the paramount aim of the countries that pushed for progress on agriculture in the UR. As we have seen (Chapter 6), the exclusion of agriculture from the general ban on export subsidies in the 'old' GATT was formulated so vaguely, and interpreted so timidly by panels, that there was practically no effective limit to the extent that countries subsidized their agricultural exports if domestic price support policies appeared to require such subsidies. It took the exporting countries, above all the US and the Cairns Group, a long time in the negotiations to convince the major user of export subsidies, the EC, that without an explicit discipline on this category of measures there would not be a UR agreement (see Chapter 7). Eventually the EC gave in, and the agreement reached on export subsidies in agriculture is both reasonably stringent and likely to be the most practically effective element in the Agreement for the years immediately following the conclusion of the UR.

Agreement Provisions

As in the case of market access, the basic rule the Agreement includes on export subsidies is very short: 'Each Member undertakes not to provide export subsidies otherwise than in conformity with this Agreement and with the commitments as specified in that Member's Schedule' (Article 8). In addition, the Agreement says that 'a Member shall not provide export subsidies . . . in excess of the budgetary outlay and quantity commitment levels specified' in its Schedule (Article 3(3)), and that the export subsidy commitments in each Member's Schedule are 'an integral part of the GATT 1994' (Article 3(1)).[27] These few sentences effectively supersede all of the GATT Article XVI:3 and Article 10 of the Subsidies Code. No more is there anything like the 'equitable share', and all language referring to 'special factors' has gone. There is no need to debate what a 'previous representative period' might be, or how to determine 'displacement' or 'price undercutting'. If one wants to know how far an individual country can go in subsidizing its agricultural exports, one turns to its Schedule and finds precise numbers regarding its commitments.

Even more important in the longer run is one other short phrase in the Agreement which says that countries 'shall not provide . . . [export] subsidies in respect of any agricultural product not specified in . . . its Schedule' (Article 3(3)). Contrary to the 'old' GATT this means, in effect, that export subsidies are now prohibited in agriculture, except where indicated in countries' Schedules.

The remaining provisions on export subsidies in the Agreement essentially aim at making these two central provisions watertight, and at creating some flexibility to shift commitments marginally between years. An important element of this tightening is the list of export subsidies falling under reduction commitments (Article 9(1)). The list of agricultural export subsidies in the Agreement contains a number of subsidy practices typical in agriculture, such as stock disposal at prices below the domestic market price; subsidies financed through a levy on the product concerned or one of its agricultural inputs (which would include so-called 'producer-financed' export subsidies); and subsidies for reducing the costs of marketing exports.[28] Article 10(1) says that 'export subsidies not listed in Article 9(1) . . . shall not be applied in a manner which results in, or which threatens to lead to, circumvention of export subsidy commitments'. Moreover, if a country claims that it has not subsidized exports of quantities in excess of the reduction commitment, it 'must establish that no export subsidy, whether listed in Article 9 or not, has been granted' on the excess quantity. Export credit benefits, a notable absentee from the list of Article 9(1), are designated as an item for further negotiations (Article 10(2)). There is a list of disciplines agreed for food aid (Article 10(4)): it is not to be tied to commercial exports, must be in accordance with the FAO *Principles of Surplus Disposal and Consultative Obligations*, and to the extent possible, given totally in grant form or on terms not less concessional than agreed in the 1986 Food Aid Convention.

The ambiguity concerning subsidies on exports of processed products under the 'old' GATT (see Chapter 6) is now removed as 'subsidies on agricultural products contingent on their incorporation in exported products' are included in the list of subsidies subject to reduction commitments. The Agreement explicitly requires that the per unit subsidy on an incorporated agricultural raw material must not exceed the per unit subsidy payable on the export of the raw material as such (Article 11).

Some flexibility for coping with year-to-year market fluctuations is provided by allowing countries to exceed their commitments somewhat during years two to five of the implementation period (Article 9(2b)). However, total cumulative outlays and quantities must not be in excess of those committed, and in year six of the implementation period they must be no greater than 64 and 79 per cent of their 1986–90 base period levels.[29] It is noteworthy that the Agreement does not contain anything comparable to the Special Safeguard Provisions on the export side. If world prices drop this is not considered an excuse for exceeding the outlay constraint on export subsidies.

In a little-publicized provision designed to meet the concerns of food importing countries (Article 12), the Agreement also somewhat constrains the imposition of export embargoes or other restrictions in cases of critical shortages of foodstuffs allowed by GATT Article XI:2(a). Countries restricting their exports are called upon to give due consideration to the effects this may have on food security in importing countries, to consult with importers, and to notify the Committee on Agriculture in advance. This provision does not apply to developing countries unless they are net exporters of the foodstuff concerned.

Schedule Commitments

The commitments countries were required to make in their Schedules were a straightforward application of these Agreement provisions on export subsidies. In their offers countries had to list, for each year of the base period 1986–90, both quantities exported with subsidies and expenditure on these subsidies, for each of the six types of export subsidies listed in Article 9(1) and for their sum. The average of these annual figures then constituted the base from which reduction commitments were calculated for each year of the implementation period. Country Schedules contain the base period levels and for each year of the implementation period the 'annual and final bound commitment levels' of outlays and quantities.

As a result of the last-minute negotiations between the US and the EC in December 1993, a 'front-loading' provision was added to the Modalities which allowed countries to start their quantity reduction commitments on export subsidies at a level higher than the base level if the quantities exported in 1991–2 were higher than in the base. However, in the last year of the implementation period, quantities exported with subsidies must still not exceed 79 per cent of the 1986–90 base quantities. The conditions for invoking this provision were formulated in such a way that they accommodated problems the EC feared regarding wheat and beef, where EC exports had continued to grow after the base period and the EC had rather large stocks. As a result of front-loading, total allowed subsidized exports of the EC over the whole of the implementation period were raised by 9.1 per cent for wheat and 6.7 per cent for beef. As in the case of the rice clause in the agreement on market access and the exception for 'production-limiting programmes' from domestic support reductions (see below), these accommodations for country and product-specific problems were moulded into a general exception that all countries that also happened to meet the criteria could use.[30]

Commitments regarding export subsidies were defined for product groups rather than for disaggregated commodities. The Modalities established a list of 22 product groups, containing, for example, two groups of cereals (wheat/wheat flour and coarse grains) and four groups of dairy products (butter/butter oil, skim milk powder, cheese, other milk products). Country Schedules contain lists of products at the six-digit HS level indicating which product belongs to which product group. By implication, any product not included in that list was not exported with subsidies in the base period, and therefore must not receive export subsidies in future. As far as export subsidies on agricultural products incorporated in exports of processed products are concerned, no quantity commitment was required, and the outlay commitment for such products comes as one sum aggregated over all incorporated products.[31]

The Modalities (paragraph 12) also suggested the possibility of negotiating commitments regarding exports to individual or regional markets. Where such commitments had been agreed they could have been specified in the Schedules. It appears, though, that this has not been done. On the other hand, bilateral understandings of this nature continue to exist, such as the 'Andriessen commitment' regarding a limit to subsidized EC beef exports to the Pacific rim,[32] and new ones are said to have been negotiated during the UR. As they have not been included in the Schedules these bilateral understandings remain outside the realm of the GATT/WTO.

Since countries had to report their base period export subsidies in their Schedules (in order to be allowed to subsidize exports in future), there is now a statistical base from which one can assess the extent to which world trade in agricultural products occurred under export subsidies in the past. As shown in Table 8.5, subsidized exports covered a large share of world trade in several major agricultural products during the base period. As a consequence, the required reductions of subsidized exports will also be large relative to total market size. Commitments to reduce the quantities of subsidized exports do not mean that actual exports from the countries concerned will fall by the respective percentages, as countries can continue to export, or even expand exports without subsidies. However, the elimination of subsidized exports in the magnitudes indicated in Table 8.5 is likely to be an important factor for market conditions in the years to come.

Implementation Issues

In the area of market access, the Agreement places direct constraints on the types and the levels of policies that countries can use. When it comes to export subsidies, the Agreement constrains the effects of policies, rather than policy instruments. How a country manages to reduce its export subsidies will depend on domestic policy decisions. As long as they allow the country concerned to meet its commitments under the Agreement they are equivalent from the perspective of the GATT, and uncontroversial in this regard. The domestic implications may or may not be in line with the underlying philosophy of an Agreement which, as the *chapeau* says, aims to foster a 'market-oriented agricultural trading system'.

Table 8.5   Subsidized Exports and Aggregate Reduction Commitments
(Selected Commodities)

	Total Volume of Subsidized Exports in the Base Period[a]		Reduction of Subsidized Exports Between Base Period and 2000	
	Million metric tons	Proportion of World Exports[b] 1992–3 (%)	Million metric tons	Proportion of World Exports 1992–93 (%)
Wheat	61.45	59.7	21.09	20.5
Coarse grains	21.24	23.0	4.98	5.4
Beef	1.75	38.4	0.48	10.6
Butter	0.64	90.1	0.15	21.6
Cheese	0.60	64.4	0.17	18.4
Milk powder	0.61	59.0	0.15	14.7

[a] Base period is 1986–90 or 1991–2, whichever is higher (because of front-loading).
[b] World exports exclude intra-EC trade.

Sources: GATT Secretariat (1994), FAO (1995).

This is most obvious in the case of the quantity commitments on export subsidies. One would expect the commitments regarding quantities of subsidized exports to be more binding than the outlay commitments. World prices have tended to increase since the base period, and major subsidizing exporters (such as the EC) have already cut their domestic support prices since that time. Consequently, reducing the quantities of subsidized exports is likely to be the most demanding policy task under the export subsidy commitments. There are obviously many different ways countries can achieve this, and it is up to governments to decide which way to go. The most economically efficient approach would be to reduce the level of price support until export availability does not exceed the quantity commitment. However, for political reasons, there may also be attractions in using the many different forms of domestic supply control which have been tried in agricultural policies for decades. In particular, where supply controls are already in place, the tightening of these supply restrictions may turn out to be politically easier than the reduction of price support. From the point of view of moving towards more market-oriented and economically rational agricultural policies, it would be unfortunate if the Agreement should result in new or more stringent domestic supply controls.

A movement towards more supply controls is not inevitable. There is one completely opposite type of policy response which can be considered if a country has difficulties with meeting its commitments in the area of export subsidies. Indeed, the most effective way of avoiding all the difficulties which may be inherent in observing the commitments regarding export competition is to eliminate export subsidies completely, by reducing the domestic support price to the level of the world price, or by giving up price support entirely. Of course, political difficulties would have to be faced if price supports were drastically reduced. However, the benefits of this option are also substantial. When a country gives up export subsidization altogether it is free to export as much as it wants. There would be no need for further supply control, and supply controls introduced earlier can be lifted – no doubt with the approval of many farmers. There would also be less reason to be concerned about future productivity growth leading to problems with the constraints. And if farm income were to suffer too much under this strategy, there would be the option of providing compensatory payments as long as they were consistent with the Agreement provisions and country commitments regarding domestic support. If these payments were to take the form of 'green box' measures, they would both be in line with GATT commitments and constitute a welcome step in the direction of 'decoupling' income support for farmers from their production decisions.

On the other hand, since the domestic support provisions of the Agreement may not be strongly binding in many cases (see below), there may be the possibility of substituting straightforward output-related deficiency payments (which do not fall in the 'green box') for export subsidies. Were this to happen, direct export subsidies would have essentially been replaced by indirect export subsidies.[33] As long as there is sufficient 'water' in the total AMS constraint, such policy substitution could help to meet the outlay constraint on export subsidies without reducing export volume. It could conceivably also be used to get around the quantity constraint on export subsidies. The domestic market price could be allowed to drop all the way to the world market level so that export subsidies were no longer necessary and the quantity restraint became void. Domestic budget concerns, however, might limit the inclination to engage in this form of policy substitution.

In implementing the provisions of the Agreement and the country commitments regarding export subsidies, a number of contentious issues may arise in the future. For example, in some cases it may not be easy to determine outlays on export subsidies because governments may be using accounting procedures that make it difficult to trace the relevant financial flows. There may also be cases where it is difficult to determine whether a given policy amounts to an export subsidy in economic terms. For example, a country may have a production quota in place which limits the quantity eligible to receive the domestic support price, which may be well above the international market price. The quota may be fixed such that the resulting volume of exports is in line with the export subsidy commitments of the country concerned. At the same time this country may allow its farmers to produce beyond the quota volume, so long as they sell the extra output internationally at the prevailing world market price. With this policy structure, no subsidies are granted on exports resulting from above-quota output. In particular, none of the subsidy practices listed in Article 9(1) is involved, and it would also be difficult to argue in a legal sense that the policy concerned is used to circumvent the export subsidy commitment. Hence, in a formal sense the Agreement is respected. However, as farmers in the country concerned can market a large share of their output at the high domestic price (and, possibly, in subsidized exports within the constraints of the export subsidy commitments), their fixed costs are covered. They can therefore sell additional output at the world market price if that price covers at least their marginal costs. Under free trade, farmers in that country might not be able to sell on the world market, if the world market price were below their average costs. In such a case, can the coverage of fixed cost through the high domestic support price be said to be a form of indirect export subsidy inconsistent with the spirit, if not the letter, of the Agreement?[34]

More serious may be the fact that all the export subsidy provisions in the Agreement relate only to the aggregate volume of exports and the aggregate outlay on export subsidies for the product groups concerned. There is no specific provision in the Agreement which relates to export sales on individual markets. When it comes to displacement of other exporters' shipments to individual export markets or to price undercutting (if there is such a thing), in principle the general rules of the GATT 1994 and the Agreement on Subsidies and Countervailing Measures (in particular Articles 5 and 6 of that Agreement) could apply, and so there might not be a need for specific rules in the Agreement. However, agricultural export subsidies that are in conformity with Schedule commitments are protected against actions under these more general rules as a result of the Due Restraint provisions (and as stated explicitly in Articles 5 and 6 of the Subsidies Agreement). Along similar lines, as export subsidy commitments in the Schedules are expressed in terms of product aggregates, rather than at tariff line level, issues may arise if a country concentrates its export subsidies on only a few individual products and if these are products of particular interest to individual exporters. Much will depend on how countries allocate the limited 'rights' to export with subsidies to individual products and exporting companies.

A clause in the provisions regarding the review process (Article 18) may also prove contentious. In it, Members have agreed 'to consult annually in the Committee on Agriculture with respect to their participation in the normal growth of world trade in agricultural products within the framework of the commitments on export subsidies under this Agreement'. This clause, born out of dissatisfaction in some parts of the EC with the need to place strict limits on subsidized exports, is open to many different interpretations.

4   DOMESTIC SUPPORT

The domestic support provisions are probably the most innovative feature of the Agreement. They go far beyond the traditional GATT approach of dealing mainly with trade policies and having no more than rather general rules regarding domestic subsidies. Indeed, right from the start of the UR negotiations it was one aim of the agricultural talks to do better than the limited achievements of earlier rounds by taking the close link between domestic agricultural policies and international trade explicitly into account. The provisions regarding domestic support are the most obvious expression of this important element in the negotiations.

Agreement Provisions

The starting point of all Agreement provisions on domestic support is the base level of AMS calculated, for each country and product, for the period 1986 to 1990. The Modalities described the method of calculation in detail. Three elements had to be included in the calculation: market price support, 'non-exempt direct payments', and other subsidies not exempted from reduction commitment. Market price support had to be calculated on the basis of the gap between the external reference price (that is the country-specific world market price) and the applied domestic administered price, multiplied by the quantity of production eligible to receive that administered price. Non-exempt direct payments and other subsidies that were dependent on a price gap were to be calculated using that price gap multiplied by the volume of production concerned, or using budgetary outlays. Support at both national and sub-national level had to be included. Levies or fees specific to agriculture could be deducted. Support which was not product specific had to be added to the total for individual products. Where products received market price support but where the price-gap calculation was 'not practicable', equivalent measurements of support were required. Support exempt from reduction commitments, and not to be included in AMS calculation, comes in three forms, *de minimis*, 'green box', and 'blue box'. *De minimis* support is that which does not, on aggregate for all policies to be considered, exceed 5 per cent of the value of production, either of individual products (for product-specific support) or of the total value of agricultural production (for non-product-specific support).[35] Green box policies are those considered to have no, or only minimal, trade-distortion effects or effects on production (see below). The base 'Total AMS' calculated in this way, and the data used, had to be included in Part IV of the country Schedules. Credit was allowed 'in respect of actions undertaken since the year 1986'. In effect that meant that countries could use as their base Total AMS the 1986 value if that were higher than the 1986–8 average. From this base Total AMS countries had to derive their 'Annual and Final Bound Commitment Levels' for each year of the implementation period, by reducing the base level by 20 per cent by the year 2000, in equal annual steps. These annual AMS levels, included in the Schedules, represent the commitments under the domestic support provisions of the Agreement.

For each year of the implementation period, the Agreement now requires countries to calculate their 'Current Total AMS'. The calculation method, described in Annex 3 of the Agreement, is the same as that for the base AMS, though current domestic prices, production quantities, and budgetary outlays are used in this case. It is, however, important to note that the external prices to be used for measuring price gaps in the calculation of current AMS are the same external prices that were used for the base period ('fixed external reference prices'). The Current Total AMS is then compared with the AMS commitment in the Schedule, and a country is considered to be in compliance with the Agreement provisions if its Current Total AMS does not exceed the Schedule commitment. For this purpose, only the aggregate AMS (that is not the AMS by product) is relevant.

Under the green box provisions, policies 'with no, or at most minimal, trade distortion effects or effects on production' have been exempted from reduction commitments, and hence these measures are also not included in the AMS calculation. The green box is defined in both a general form and in terms of an illustrative list of eligible policies, which includes measures such as advisory services, domestic food aid, decoupled income support, income insurance and safety-net programmes, set-aside payments (if land is retired for a minimum of three years), regional and environmental aids, and those that encourage early retirement (Annex 2 of the Agreement). Policies on this list need to meet specific criteria defined in that list. However, it is important to note that, in addition, they also have to meet the general condition that they should have no, or at most minimal, production-distorting effects.

As a result of the Blair House Accord between the United States and the EC, the 'blue box' exemption was agreed, with the result that both US deficiency payments and the new compensation payments under the reformed Common Agricultural Policy of the EC do not need to be included in the AMS calculation and the reduction commitment. The wording chosen for this purpose exempts 'direct payments under production-limiting programs' if they are made on the basis of fixed area and yield (or number of head for livestock) or on a maximum of 85 per cent of base level of production (Article 6:5). Though this provision obviously targets the two country-specific cases mentioned, it is general and can therefore be invoked by any country for any policy meeting the criteria specified.

The Concept of Domestic Support

The domestic support provisions of the Agreement have established a new concept in GATT rules for agriculture. The fact that there are binding quantitative commitments in three areas, that is market access, export subsidization, and domestic support, may appear to suggest that there are specific bindings on border measures on the one hand (tariffs and export subsidies), and separate specific bindings on domestic subsidies on the other. However, that is not really the case. One needs to interpret both the 'domestic' and the 'support' very carefully to see why. One way to understand the rather specific nature of the domestic support commitments is to look at the differences between the PSE and the AMS.[36]

For a product with a domestic administered support price (e.g. an intervention price for government buying activities, or a target price for deficiency payments), there is no fundamental difference between the PSE and the AMS. In both cases the level of support would be calculated as the aggregate of market price support (difference between domestic and world market price), multiplied by the volume of production, and the expenditure on (non-output price) domestic subsidies.[37] There is, however, a significant difference between the two measures where an administered price does not exist. In this case, too, the domestic producer price may well be above the world market price, as a result of import duties or, if the country is an exporter, export subsidies. In measuring the PSE this difference between the domestic producer price and the international market price would be included, along with any domestic subsidies that may be granted at the same time. In calculating the AMS, however, the price gap would not be taken into account because there is no administered price, and only the domestic subsidy would be included. The PSE is defined by the effects of all policies, whether operating internally or at the border. The AMS, on the other hand, is defined by the effect of a subset of policies, excluding those that operate at the border. The AMS measures the effect of domestic policies in the sense that it does not include support that is provided only through border measures, and it measures support in the sense that it not only includes domestic subsidies but also price support if provided through an administered price.

To some extent, the domestic support provisions can, in theory, still be thought of as being the most important element in the Agreement for the 'typical products' under traditional agricultural policies in industrialized countries, where administered prices exist. The AMS commitments establish a comprehensive discipline for all agricultural policies affecting these products. If anything 'goes wrong', in the sense that governments find ways to evade some of the more specific disciplines, there is still the overarching constraint on the aggregate effect of what they can do in their agricultural policies. In practice, however, for several reasons, even in these cases the provisions on domestic support are likely to be less constraining in most cases than other provisions in the Agreement.

First, with the agreement to accept specific commitments in the areas of market access and export competition, much weight is taken off the AMS approach. Second, since the AMS commitments are not product specific but sector-wide ('Total' AMS), there is the possibility of shifting support among different products. At the same time, for many products, the AMS will be driven down by tariffication and export subsidy commitments more than by what the AMS commitment *per se* would have required. The 'slack' in the total AMS created in this way can be used for other products. Hence the sector-wide nature of the AMS commitment has greatly increased the scope for policy freedom, and reduced the impact of the AMS bindings. Third, the exclusion of price support in cases where no administered price exists results in rather different treatment of not too different cases, and opens up the possibility of alleviating the AMS commitments through policy re-instrumentation (that is by replacing administered prices with border policies, though bound tariffs and export subsidy constraints make this difficult). Fourth, it has to be remembered that the domestic support provisions do not really establish direct commitments regarding domestic subsidies in the narrow sense. Fifth, a relatively large set of domestic subsidies is exempted from the reduction commitments as being relatively trade-neutral, and hence there is some scope for governments in choosing policy instruments.

Given the fact that all border measures are covered by specific binding commitments and that a domestic support price cannot be implemented without appropriate border measures, the only group of measures left to be effectively bound by the AMS is domestic subsidies. In that sense one could think of the AMS commitment as the domestic subsidy commitment, even though the AMS calculated according to the Agreement includes more than just domestic subsidies. On the other hand, looked at from that perspective there is some redundancy in the AMS approach adopted in the Agreement. It might have sufficed to bind domestic subsidies proper by agreeing on maximum budget outlays, as was considered at one stage during the negotiations. Now that the comprehensive AMS approach has been adopted, for products with administered prices border measures are constrained by both the commitments specific to them *and* by the AMS commitment.[38] This redundancy might be welcome in the sense that it may be better to have two checks rather than one. However, the flip side of the same coin is that the AMS is less binding on domestic subsidies as narrowly defined.

An important element of the domestic support provisions is the long list of policies that are exempted from reduction commitments. The definition of these policies was hotly debated in the negotiations. The amount of text used to describe the largest category, that is green box policies, five pages in Annex 2 of the Agreement, is indicative of the detail in which the matter was addressed. The long list of green box policies looks worrying at first glance because it exempts so many policy measures from the domestic support discipline. However, the length of that list is more indicative of the many things governments do than a reason to be concerned about trade distortions that may arise from these exemptions. The general conditions (no, or at most minimal, trade-distortion effects or effects on production) apply to all policies for which exemption is claimed.[39] Hence, all green box policies must meet the general criteria *plus* the policy-specific conditions spelled out for each of them. In principle one could have left the conditions for exemption from reduction commitments in the general form of the non-distortion criterion. However, it would have then been a matter of developing case law as to which measures qualify under these conditions. Negotiators considered it advisable to be more specific right from the start, and thus a long list of examples of policies was drawn up. Independent of the assessment of any particular policy included in the list, this approach is useful in the sense that it provides guidance as to how to draw the borderline between green and not-so-green policies. After all, there are very few, if any, policies that really have no effect at all on production and hence on trade, and it is therefore a subjective matter to define what a 'minimal' distortion effect is when it comes to determining whether or not a given programme is eligible for exemption.

In general the programmes included in Annex 2 appear to have two characteristics. First, they are much less distorting than traditional forms of agricultural support provided through market, price, and trade policies. Second, it would be difficult to imagine that governments would be prepared to give up these relatively innocent policies in trade negotiations. This is not to say that no trade impacts can be expected to follow from such policies. Indeed, some of the measures included, in particular research, general services, structural adjustment assistance provided through investment aids, and payments under regional assistance programmes, may well stimulate production and hence result in higher exports or lower imports. Moreover, there is not, unfortunately, a general requirement that green box measures must not be commodity specific. However, many of the green policies fall under the heading of the provision of public goods or the pursuance of internal development and growth policies. As such, they are unlikely to be controlled by trade policy. Moreover, in net exporting countries the export subsidy commitments are an additional cap on the effects of such policies. In any case, the examples of policies included in the green list are useful at least in the sense that they indicate to policy-makers in which directions they should reinstrument their agricultural policies.

On the other hand, the exemption of some payments under production-limiting programmes (blue box), if they meet certain conditions, is a true departure from the aim of moving in the direction of less distorted agricultural markets. The payments referred to in this provision are not limited to compensation for not producing a given share of base period production, but cover all payments made under programmes that, among others, require farmers to limit their production. Contrary to the definition of decoupled income support in Annex 2, there is not the condition that 'no production shall be required in order to receive such payments'. Not only can governments require farmers to produce in order to be eligible for these exempted payments, they can also make the size of payments directly dependent on the volume of production, so long as it is on no more than 85 per cent of the production in the base period. Hence payments allowed under this provision can well have a production-stimulating effect, and there is no guarantee that this effect is less than that of the production limitation which also must be an element of the programme. As a result, governments have wide scope for designing policies that may be brought under this provision, and the trade effects of some of these policies may be undesirable. It must also be remembered, however, that the 'peace clause' may impose some constraint on policies under production-limiting programmes (and *de minimis* measures), since they are exempt from GATT challenges only if the level of support provided by them does not exceed that granted in 1992. Moreover, under the peace clause provision the AMS measure is product specific.

In conclusion, the provisions on domestic support, though an important conceptual element of the Agreement, are unlikely to be very constraining in the short run. In the longer run, the fact that domestic support is bound in nominal terms may mean that, with some inflation, the level of allowed domestic support may become more binding.[40] Moreover, the list of green box policies in Annex 2 is a step forward in the sense that it indicates in which direction agricultural policies should be going in future, in order to achieve domestically what they are supposed to achieve without negative international effects.

5 SANITARY AND PHYTOSANITARY MEASURES[41]

The Agreement on Sanitary and Phytosanitary Measures is separate from the Agreement on Agriculture, but it has obvious and close links with it, not only because both agreements relate to the same category of products, but also in a more substantive way. Commitments to reduce economic barriers to trade may reinforce tendencies to use technical standards to provide continued protection to domestic producers.

In agriculture, the most important technical standards are sanitary and phytosanitary measures safeguarding human, animal, and plant life and health. It is therefore important that the application of sanitary and phytosanitary measures (SPS) is disciplined so that they cannot be utilized as substitutes for the economic measures which are being liberalized under the Agreement. In addition to guarding against such inappropriate use of SPS, there were also many issues relating to SPS in their own right which needed to be better regulated at the international level. In particular, it was important to make sure that these measures can achieve their important aims in the protection of human, animal, and plant life and health, while not resulting in economic waste by restricting the international division of labour to more than an unavoidable extent. Countries have always disagreed on the appropriate level of sanitary and phytosanitary protection, and there were a growing number of cases where given SPS were considered, by the exporting countries, to be unjustified barriers to trade. Thus it was considered imperative that progress be made in the UR towards establishing more concrete and more easily enforceable rules for SPS. The Agreement on SPS is a step in this direction, and it has the potential of proving as important in its area as the Agreement on Agriculture is in the realm of economic measures. It is therefore appropriate to think of the new rules regarding SPS as the fourth pillar of the UR achievements in agricultural trade, complementing the new rules and commitments in the three areas of market access, export competition, and domestic support.

Under the GATT 1947, SPS were governed by Article XX(b) which stipulates that

> [s]ubject to the requirement that such measures are not applied in a manner which would constitute a means of arbitrary or unjustifiable discrimination between countries where the same conditions prevail, or a disguised restriction on international trade, nothing in this Agreement shall be construed to prevent the adoption or enforcement by any contracting party of measures . . . (b) necessary to protect human, animal or plant life or health. . . .

However, there was no definition of the criteria to judge the 'necessity' of a sanitary and phytosanitary measure and whether it constituted 'unjustifiable discrimination', and there was also no specific procedure for settling disputes on such measures. In the Tokyo Round an attempt was made to improve GATT rules on technical standards, including SPS, through the Standards Code (see Chapter 5). The strengthened rules of the Standards Code – which included emphasis on harmonization of standards through international bodies, notification requirements, and a specific procedure for settling disputes regarding technical standards – removed some of the deficiencies of Article XX and created more transparency. But a number of issues remained unresolved, among others the treatment of standards relating to production and processing methods (as opposed to product standards). Moreover, only 37 countries had signed the Code. A number of concrete cases regarding SPS could not be settled satisfactorily under the Standards Code, for example the ban on animal growth hormones introduced by the EC in 1988.[42] Hence it appeared necessary to agree on better and more operational rules in the UR. As sanitary and phytosanitary issues in agriculture are of a rather specific nature, their regulation was not left to the strengthened Agreement on Technical Barriers to Trade which was also negotiated in the UR. Instead, a separate SPS Agreement was designed.

In line with the GATT Article XX(b) the SPS Agreement confirms the right of countries 'to take sanitary and phytosanitary measures necessary to the protection of human, animal or plant life or health'. However, a number of new criteria are defined which have to be met when it comes to adopting or maintaining such measures. If SPS conform to these criteria, they are considered to be in accordance with Article XX(b). The cornerstones of the SPS Agreement are the concepts of 'scientific evidence', 'risk assessment', 'appropriate level of protection', and 'equivalence'. An important role is foreseen for 'harmonization'. Notification requirements are defined to enhance transparency, and specific provisions for dispute settlement are intended to resolve disagreements which arise. The SPS Agreement covers regulations concerning product characteristics, but now also explicitly refers to measures regarding processes and production methods (PPM), testing, certification, and other related measures.

'Scientific evidence' is introduced as a new condition for the necessity of a sanitary and phytosanitary measure, and countries are going to have to meet this test. Members must apply any such measure 'only to the extent necessary to protect human, animal or plant life or health' and ensure that 'it is based on scientific principles and is not maintained without sufficient scientific evidence'. No definition or standard of scientific evidence is identified, nor could there have been for there are many instances where scientists disagree. However, where an overwhelming majority of scientific analyses have come to the conclusion that a given sanitary and phytosanitary measure is not necessary to protect life or health, it is now more difficult for a government to maintain that measure.

Closely related is a provision which may guard against the argument that even the slightest remaining doubt about health risks may be sufficient to justify protective measures. This is the requirement to base SPS on a 'risk assessment'. This interesting concept which is new to the GATT but not to other institutions (Wirth, 1994), requires Members to 'take into account . . . relevant economic factors' such as the potential damage in terms of loss of production or sales resulting from a pest or disease, the cost of control or eradication of that pest or disease, as well as the relative cost effectiveness of alternative approaches to limiting risks. When determining the appropriate level of SPS protection on the basis of such a risk assessment (in an annex to the Agreement also referred to as the 'acceptable level of risk'), countries must now bear in mind 'the objective of minimizing negative trade effects'. In other words, the SPS Agreement makes an attempt to guard against the unfettered use of SPS irrespective of the health risks involved and of the negative economic and trade implications which excessive levels of SPS protection might have.[43] However, a loophole is provided: 'where relevant scientific evidence is insufficient' countries may adopt measures 'provisionally'. In such cases, though, countries 'shall seek to obtain the additional information necessary for a more objective assessment of the risk' involved and review the measure concerned 'within a reasonable period of time'.

Problems have often resulted where different countries use different measures to guard against the same sanitary or phytosanitary risk. In such cases the SPS Agreement now establishes the principle of equivalence of different measures. Where the exporting country's measures, even if different from those of the importing country, achieve an 'appropriate level of . . . protection', importing countries shall accept them as equivalent to their own regulations. The principle of equivalence now explicitly adopted means that the PPM used in the exporting country cannot be questioned by the importing country as long as they achieve the required product characteristics. This is a major step forward since many SPS come in the form of PPM. As a result, disagreement over importing countries' rights to require that exporting countries adopt certain PPM was an important source of conflict before the SPS Agreement entered into force.

Another most promising element in the Agreement is its emphasis on international 'harmonization' of SPS. The Agreement suggests that in order to harmonize their regulations 'on as wide a basis as possible' countries 'shall base their sanitary or phytosanitary measures on international standards, guidelines or recommendations, where they exist'.

The incentive for doing so is the Agreement provision that measures conforming to international standards are automatically 'presumed to be consistent with the relevant provisions of . . . [the] Agreement and of the GATT 1994'. In other words, where a country uses internationally accepted measures it is safe from challenge by its trading partners. In this context the Agreement stipulates that 'Members shall play a full part within the limits of their resources in the relevant international organizations'. The international institutions specifically mentioned are the Codex Alimentarius Commission, the International Office of Epizootics and the International Plant Protection Convention. Where a country applies a measure resulting in a higher level of protection than achieved by measures based on international standards, either there must be a scientific justification or the measure must be based on an appropriate risk assessment. This requirement is not exactly clarified by a note which says that 'there is a scientific justification if, on the basis of an examination and evaluation of available scientific information . . ., a Member determines that the relevant international standards . . . are not sufficient to achieve its appropriate level of protection'.

When it comes to disagreement on SPS, the general procedures on consultations and dispute settlement of the GATT apply, as amended by the WTO agreement. More specifically, though, when scientific or technical issues are involved, the panel dealing with the case can now seek advice from experts. To this end, the panel may 'establish an advisory technical experts group, or consult with the relevant international organizations'. *Ad hoc* consultations or negotiations on specific sanitary or phytosanitary issues shall be pursued in the newly formed Committee on Sanitary and Phytosanitary Measures, which will oversee the implementation of the Agreement. The tasks of that Committee also include the monitoring of the process of international harmonization and of the use of international standards. In addition, the Agreement has rules on the publication of regulations, enquiry points and notification procedures. It also requires central governments to ensure (or use their best endeavours to ensure) that authorities at lower levels and nongovernmental bodies within their territories do not undermine the provisions of the Agreement, and it contains detailed rules for control, inspection, and approval procedures.

As a consequence of the nature of measures dealt with, the character of the Agreement on SPS is rather different in many regards from the Agreement on Agriculture. In particular, the Agreement on SPS does not specify any quantitative requirements and, more importantly, it does not (and, given the subject matter, could not) regulate SPS in any detailed way. Hence, individual countries do not commit themselves to specific changes in their SPS policies and practices. The Agreement on SPS rather establishes general guidelines for government behaviour in the area concerned. Some of these guidelines are inevitably open to interpretation. It is therefore not directly evident which measures in which countries will have to change, and in which ways. As a consequence, it is difficult to spell out the implications of this Agreement for trade flows. Much will depend on the spirit in which governments implement their SPS under the new guidelines. The real test of the Agreement will be in the way disputes are settled under it.

Where harmonization is achieved, that is where national SPS are based on standards agreed in the relevant international institutions, disputes should not arise. On the other hand, harmonization at the international level may not always be achievable or even appropriate, and the Agreement therefore also offers the alternative option of equivalence. Indeed, in practice, widespread application of the equivalence principle may in the longer run result in growing harmonization, as trading partners learn more about each other's measures and find out which measures are most appropriate in which cases. Where a country does not rely on harmonization or equivalence, but rather insists on its own domestic standards, it has to comply with a number of requirements. It has to make sure that its measures do not discriminate between countries where identical or similar conditions prevail, and it must not apply them as a disguised restriction on international trade. These requirements are fundamentally those that always existed under the GATT Article XX, except that they are now somewhat more demanding.[44] Additionally, there is now also the obligation to ensure that the measures concerned are consistent with scientific evidence and that they are based on an appropriate risk assessment. These additional requirements are completely new in the GATT, and governments will have to develop suitable procedures for monitoring them. There is no doubt that these new requirements will demand adjustments in many countries' procedures for introducing and implementing sanitary and phytosanitary measures. If governments honour the spirit and the letter of the SPS Agreement it will turn out to be an important adjunct to the Agreement on Agriculture.

6 REMAINING DIFFERENCES BETWEEN AGRICULTURE AND INDUSTRY

The UR Agreement on Agriculture takes a large step towards integrating trade in farm and food products fully into the GATT. It establishes binding and operationally effective rules and commitments for agricultural policies in participating countries. By doing so it begins to close the gap which has existed throughout the history of the GATT between industry and agriculture. However, some important differences between these sectors remain. It would have taken more than one round of multilateral trade negotiations to eliminate all the special features which have characterized agriculture in the GATT for nearly half a century.

The most prominent remaining difference between agriculture and industry is the continued legitimacy of export subsidies in agriculture. In industry, export subsidies have been explicitly prohibited since 1955 by the GATT Article XVI:4. This prohibition has been reinforced by Article 3 of the UR Agreement on Subsidies and Countervailing Measures (Subsidies Agreement). In agriculture, export subsidies were always[45] legitimate within the 'equitable share' provision of the GATT Article XVI:3. The UR Agreement continues this tradition: export subsidies are still legitimate in agriculture, where countries had used them during the base period. However, the extent to which they can be used is now defined, in very specific quantitative terms, for each individual country, product category, and year, in the Schedules. This can be interpreted as giving a much more concrete meaning to the GATT provision regarding the 'equitable share', though the quantities which can now be exported with subsidies have never been declared 'equitable' in any sense. Moreover, new subsidies on exports of products that were not subsidized in the past are now prohibited in agriculture.

In the area of market access, it could be argued that there are now more explicit differences between agriculture and industry than used to be the case before the UR. The major differences are the Special Treatment Provisions ('rice clause'), the Special Safeguard Provisions, and the provisions regarding minimum access. All these special provisions for agriculture did not exist in the GATT 1947, and they have no equivalent in industry. However, these special agricultural provisions simply acknowledge the fact that import barriers in agriculture have always been special. And as in the case of export subsidies, the new agricultural rules, though making agriculture more specific in a technical sense, are a step towards removing the differences between agriculture and industry. Indeed, the big step made in that direction is tariffication, and all of the new agriculture-specific rules are adjuncts to that process. Also, they are temporary, either explicitly or implicitly. The rice clause allows tariffication of NTM to proceed in general, without a few countries blocking the whole process. The Special Safeguard Provisions remove some of the fears governments and farmers had in the face of tariffication, and thereby made it possible for tariffication to be accepted. Both clauses are supposed to expire at some stage, though not at a fixed date. Minimum access commitments are thought to speed up the process of market opening, and they will become irrelevant once tariffs have been sufficiently reduced so as to eliminate the difference between 'normal' tariffs and the lower tariffs charged on TRQ under minimum access commitments.

There is now one difference in the area of market access which makes agriculture appear progressive relative to industry in the GATT. In agriculture, essentially no unbound tariffs and non-tariff measures remain after the UR, with the very few exceptions resulting from the rice clause. In trade in manufactures, this state of affairs has not yet been reached. Hence it can be argued that tariffication has set a good example, which should be followed by other sectors in future rounds of negotiations.

In the area of domestic subsidies, too, agriculture now looks more special in legal terms than it used to do before the UR. In the GATT 1947, rules on domestic subsidies (which were weak anyhow) did not distinguish between primary products and manufactures. The UR Agreement now makes an explicit distinction. Rules in the Subsidies Agreement on actionable subsidies, against which countervailing measures can be applied, are explicitly said not to apply to agriculture, under certain conditions. Instead, the Agreement has its own rules regarding domestic subsidies. Hence agriculture is now an explicitly special case in the area of domestic subsidies.

In industry, the border line between 'green' (that is non-actionable) and 'amber' (that is allowed but actionable) subsidies, defined in the UR Subsidies Agreement, is specificity (to an enterprise or industry). All agricultural subsidies falling into the agricultural green box are specific in that sense, and would therefore be amber in industry. However, the Due Restraint provisions in the Agreement on Agriculture explicitly make these green agricultural subsidies non-actionable, and therefore also green in the sense of the Subsidies Agreement. In agriculture, amber subsidies (those with a more than minimally trade-distorting effect) are 'half-actionable', and less actionable than in industry as the 'peace clause' calls for due restraint in initiating countervailing duty investigations, and exempts them from other GATT challenges so long as (product-specific) support does not exceed the level decided in 1992. As a *quid pro quo*, amber subsidies in agriculture have to be reduced under the Domestic Support commitments.[46] In other words, green and amber do not really have the same meaning in agriculture and industry, either in terms of their definition, or in terms of the rules applying to them.

Within the group of amber agricultural subsidies, there is, however, also a subgroup of domestic subsidies that do not need to be reduced. This subgroup comprises the *de minimis* cases and direct payments under 'production-limiting programmes', as well as certain subsidies in developing countries. To complete the spectrum of traffic light colours,[47] there are also 'red' subsidies, that is subsidies that are prohibited. In industry, export subsidies are red. In agriculture only those export subsidies that did not exist in the base period are red (though one could possibly also call export subsidies in excess of those covered by Schedule commitments red).

In addition to being confusing, some of these differences between agriculture and industry in the area of domestic subsidies are less easy to justify than differences in export subsidies and market access. In the latter two areas, governments have given up some scope for policy design, and differences between agriculture and industry are in effect being reduced. The remaining differences in export subsidies and market access, thus, in a way indicate how far governments have already gone in removing the special treatment of agriculture. In the area of domestic subsidies, on the other hand, a completely new distinction has been created, by calling certain subsidies green that have not been green before. Also, there is no indication of a commitment to bring the agricultural green box under the stricter rules for green industrial subsidies in the future.

However, it can be argued that the new distinctions between agriculture and industry created in the area of domestic subsidies were a reasonable price to be paid in order to make progress towards less special rules for agriculture in the areas of export subsidies and market access. Indeed, if governments had not had the option of paying some form of direct income support to agriculture it is unlikely that they would have agreed to a reduction in market price support, and to opening their domestic markets to more international competition, and to multilateral disciplines on their export subsidy practices.

9 The Future for Agriculture in the GATT

1 ASSESSMENT OF THE POST-URUGUAY ROUND SITUATION

The new era in agricultural trade ushered in by the Uruguay Round Agreement on Agriculture and the establishment of the World Trade Organization is one of tantalizing promise. Much of this promise stems from the fact that an agreement of substance was in the end achieved. Countries at last faced squarely the troublesome set of issues of high agricultural protection and world market disruption. In the end, unlike the agreements of the Kennedy and Tokyo Rounds, negotiators did not blink. Tough decisions were made rather than postponed or fudged. Yet the promise of a liberal, rule-based structure for agriculture remains unfulfilled. The process of fully reforming the trade system for farm products will take many more years, and will try the determination of countries to stay the course. In particular, the high levels of protection which remain in agricultural markets have now been revealed in the form of the newly-bound tariffs.[1] Removing that protection will be a formidable task, but future negotiations can build on the foundations laid in the Uruguay Round.

Achievements in Agriculture

The rules governing agricultural trade have undergone a major change as a result of the Uruguay Round. The Agreement on Agriculture has extended to agriculture the tariffs-only rule for border protection, banned additional export subsidies, categorized domestic policies into those that are acceptable and those deemed trade distorting, and improved the rules on the use of sanitary and phytosanitary standards. The process of converting NTM to tariffs and the binding of those and other tariffs is clearly the most significant of these new developments. The existence of GATT-bound tariffs has the great advantage of limiting the influence of domestic lobbies on the formulation and operation of trade policies.

217

Other benefits to the multilateral system of agricultural trade should follow from the use of tariffs. The trade system as a whole is likely to gain stability under a tariff-dominated system as the burden of adjustment to shocks in global supply is spread more widely among participating economies. Non-tariff import barriers, by intention, protect importing countries from such changes in world markets. This in turn puts more of the burden on other countries to adjust stocks, consumption or production. Over time, this additional stability could help foster the dismantlement of the remaining high levels of protection, some of which have been justified in the past on grounds of defence against depressed world price levels. Tariffs could even curb the use of export subsidies in some instances, since arbitrage will limit the size of export subsidies to the level of the tariff plus transport costs. The move to tariffs will reinforce, and even lock in, the widespread movement toward private trading systems in developing countries, particularly those in Latin America, where government-issued licences have been replaced by tariffs.

Tariffication of import barriers in agriculture, as agreed in the Uruguay Round, is a quantum leap forward in the long process of bringing agriculture fully into the realm of the GATT. Indeed, in this respect agricultural trade is now ahead of trade in manufactured goods, where not all tariffs are yet bound and where non-tariff measures continue to play a significant role. Unfortunately, for the time being this fundamental change in the rules of the game has been achieved at the cost of making relatively little progress towards liberalization of market access in farm products. Reducing by 36 per cent on average both the new tariffs, with their generous cushion resulting from 'dirty tariffication', and existing tariff rates, many of which were very high in 1986–8, still leaves a considerable amount of protection. Hence the challenge for the future is to continue the process of protection removal. In part this can happen through the rigorous implementation of the current agreement, but beyond that further progress will have to wait until the agricultural 'mini-round' of negotiations about continuation of the reform process, foreseen in the Agreement to be initiated by 1999.

The rule change for agricultural trade most desired by the Cairns Group, a ban on the use of export subsidies, eluded the negotiators. Just as the unsatisfactory conditions for market access were revealed by the process of tariffication, so the full extent of the use of export subsidies in the past few years has now been quantified. The Uruguay Round made a start by curbing the use of export subsidies for agricultural products, and has all but eliminated the threat of a new 'subsidy war' in agricultural products, but the pace of reduction in the existing subsidies is somewhat disappointing. Reducing export subsidies by 36 per cent in expenditure and 21 per cent in volume from high base-period levels will by no means end the use of such subsidies.

The negotiation of constraints on the domestic agricultural policies of the contracting parties was a major innovation of the Uruguay Round. The outcome, however, is not likely to prove as effective as might have been wished. The US and the European Union (EU)[2] were able to strike the bargain at Blair House that removed the payments under their respective cereals programmes from the discipline of reduction. The EU was obliged, in effect, to keep its variable levy for grains as a result of a 'maximum duty-paid import price' provision set at the level of the current threshold price. Neither the US nor the EU is likely soon to be near the ceiling imposed on the level of its domestic support, nor on its export subsidy expenditure. In fact, a selling point of the agreement used by governments to domestic farm constituencies was that few changes would be needed in agricultural policy. Thus the GATT agreement will be most useful as a backstop, to prevent backsliding of domestic reforms in agricultural policies and to encourage those countries that are undergoing reluctant reform to stay the course.

Does this mean that the Round had no effect on domestic policies? This would be too hasty a judgement. First, many countries will be constrained in their policy decisions by the AMS limits. Those countries that are not affected are those that have reduced (amber box) transfers to farmers in the period since 1986. Thus, the countries that have begun the process of reforming their national agricultural policies will have little additional pressures arising from the limits on total support. New Zealand, Mexico, and Sweden are the leading examples of countries having undergone radical reforms, but a number of others have also scaled back support to agriculture in the last few years. Those that have yet to start this process, including many in Asia and the remaining EFTA countries, will find the AMS constraint more binding. Second, the device of the green box will provide a significant incentive for countries to reinstrument their policies to give income support to farmers in a way that is consistent with green box criteria. Countries that had been grappling with these issues on a national basis now have the guidance of international rules.[3] Over time this should improve the workings of the world market. Third, the existence of any constraint, however loose, is an important first step in international control of domestic policies. The precedent is set and the mechanism established. Further negotiations can build upon this start. The Uruguay Round offered the chance to improve trade relations between the US and the EU. The negotiations themselves proved to be a source of prolonged and intense conflict, and possibly exacerbated tensions for a time. The outcome, however, affords at least a breathing space for a more constructive relationship to develop. The inclusion of a peace clause, as described in Chapter 8, at least indicates the willingness of both the EU and the US, and of the other signatories, to try to put aside trade conflicts for a limited period.

Does this mean a period of peace for agricultural trade relations? Such an outcome is unlikely. US–EU agricultural conflicts will not disappear overnight, but they will take place within a clearer framework of rules and obligations. More likely as a scenario is a continued period of tension, with both the EU and the US trying to interpret the terms of the Agreement in its favour, at the same time accusing the other of bad faith. The prospects for trade peace depend on the extent to which these and other countries accept dispute settlement rulings that are against the interests of powerful domestic lobbies. The dispute settlement process itself has been strengthened to avoid the blocking of panel reports by the 'losing' party: it remains to be seen whether these losses will be accepted by national politicians.

Lessons from the Uruguay Round Negotiations in Agriculture

The Uruguay Round offers a number of lessons for the future of trade negotiations in agriculture. These lessons are an important aspect of the assessment of the Round itself. The negotiations should be seen as a continuation of the process started in the mid-1950s to tackle the systemic problems of agricultural trade. In earlier rounds of GATT negotiations this attempt was thwarted by the unwillingness of first the US and then the EU to subject their domestic farm programmes, with their acknowledged international ramifications, to any form of discipline in the GATT. The Uruguay Round for the first time brings the policies of the agricultural superpowers and those of other countries into a framework of rules and quantitative constraints. Thus the question as to how this was possible is itself a key to any future agenda. Similarly, to explore what has changed to allow these international rules to be agreed may answer the question as to the long-run viability of the new agricultural system. Finally, the lessons from the Round speak directly to the future of trade negotiations in agriculture and help in the attempt to look forward to the agenda for the next decade.

The first lesson is that the Uruguay Round, as with all the other Rounds, was a child of its time. It not only reflected the economic and political developments of the early 1980s but also the economic philosophies and policies of the major trading countries. The decade from the first discussions of the Round through to its completion was a remarkable period of change, perhaps unprecedented in modern history. The spread of the paradigm of economic liberalism throughout the industrial world, its adoption by many developing countries as a structural prerequisite to growth, and its ascendancy over the alternative central-planning model of economic policy, provided the precondition for a move forward in trade policy.

The inclusion of agriculture in the new trade system for a liberal world was not, of course, assured. For this to happen there had to be a change in the philosophy and practice of agricultural policy at the national level. These changes were forthcoming, at a different pace in different countries but inexorable and in a common direction. The implicit political compact between rural and urban areas – that the government would provide sheltered markets and stable prices for agricultural products in return for the stability and electoral support of rural areas – began to break down. In its place a new paradigm in agricultural policy emerged which recognized that the social and environmental objectives of supporting agriculture could be usefully separated from the manipulation of commodity markets. It was the latter that could be reduced by mutual consent at the negotiating table, to the advantage of the trade system as well as to domestic consumers and taxpayers. The Uruguay Round Agreement on Agriculture was the beneficiary of this change in agricultural strategy at the national level.

The history of the earlier GATT Rounds has demonstrated the importance of the state of world markets for the success of trade talks on agriculture. The high prices for agricultural raw materials in the early 1950s gave way to the surpluses of the second half of that decade. The cycle was repeated in the 1970s and again in the 1980s.[4] The need to off-load cereal surpluses onto world markets in times of oversupply inhibited the US from allowing limits to export subsidies in the 1950s. By the 1960s the US was beginning to look for help from others to manage the oversupply. When the EU developed into a major exporter, international agreements to manage markets became more acceptable to the US. Import access was a dominant issue when world prices were low but almost disappeared from the agenda in the Tokyo Round when world prices climbed. The drop in world market prices around the mid-1980s made a new round of negotiations on agriculture appear an imperative for many exporting countries. Higher prices on world markets in the late 1980s modified the proposals in the Uruguay Round, removing the need for short-run action to support world markets, and allowing the focus to be on longer-term changes.[5]

Over time the nature of the possible rules of agricultural trade and the constraints on national policy action have been changing. On the one hand, the negotiation of trade rules for individual agricultural commodities has declined. In the first thirty years of the GATT, negotiations were dominated by commodity-specific discussions, either in the form of requests for and offers of concessions or of commodity agreements. The Uruguay Round outcome is a set of rules which apply to all commodities, and the negotiations deliberately avoided the setting up of commodity groups. On the other hand, the commitments into which countries have entered have become more specific and are now defined in quantitative terms, culminating in the detailed schedules of tariffs and allowable export subsidy levels. General rules of a purely qualitative nature proved inadequate to contain the trade effects of national policies.[6]

A final lesson from the Uruguay Round as it relates to agriculture is that persistence sometimes pays off. Tariffication as an idea has surfaced at every major discussion of agricultural trade rules since 1947. The conditions for agreement on tariffication were apparently not fulfilled until the UR.[7] The concept of negotiating on domestic policies emerged from the Haberler Report in 1958, but was not agreed until the major countries had convinced themselves, helped by studies such as the OECD Trade Mandate, that this was an important way out of the vicious circle of subsidy and counter-subsidy. The notion of limiting the level of support had been suggested by the EU in the Kennedy Round and not adopted. It came up again in the US proposal of 1987 in the Uruguay Round. It was seen as a possibility in part because the OECD study had shown that the data problems could be overcome and in part as a way out of the commodity-specific cast of previous negotiations. In the end it proved to be less than central to the UR outcome, but could be a platform on which to build in future.

It is difficult to say whether the agricultural trade system has a 'collective memory', or whether individuals in government and intergovernmental agencies stay in place long enough to impart consistency in trade policy matters. More likely, each generation faces similar situations with a mixture of wisdom imparted from past experience, new (or recycled) ideas and analytical tools, and fresh empirical evidence. Thus the continuity and the discontinuity may reflect a balance between drawing from the past, on the principle that 'there is nothing new under the sun', and trying a new approach as, by definition, old remedies must not have worked if the problems still exist. In agriculture this means that some ideas such as tariffication were good ideas whose time eventually came, and some ideas such as commodity agreements were deemed to be failures and were abandoned in favour of export subsidy constraints. It remains to be seen whether the development of overall support and domestic policy constraints proves to be an emerging idea, generalizable to other aspects of trade policy, or one of fleeting acceptance and limited scope.

The major task for the future in agricultural trade is to continue the start made in the Uruguay Round towards a rule-based system which allows countries to develop their agricultural potential in a transparent and stable environment. This task will have to be accomplished alongside other developments in the trade system, ranging from the incorporation of new members in the GATT/WTO multilateral system to the consideration of 'new' agenda items in international negotiations and in trade rules. It will also be influenced, positively or negatively, by the development of regional trade blocs and other changes in the architecture of the trade system. Most importantly, it will depend upon the continued adoption of the paradigm of economic liberalism by developed and developing countries, and by the progressive application of that paradigm to agriculture.

2 CONTINUATION OF THE PROCESS OF AGRICULTURAL TRADE REFORM

The Mini-Round

The Uruguay Round itself has determined the next steps for the multilateral process of trade liberalization. The Agreement on Agriculture calls for talks to be initiated no later than 1999 on the continuation of the process of reform of the trade system for farm products. The responsibility for review of the implementation of the UR Agreement on Agriculture rests with the newly formed Committee on Agriculture (CoA).[8] This Committee will carry out the review on an ongoing basis in its regular meetings. The Agreement lays out 'the long-term objective of substantial, progressive reductions in support and protection resulting in fundamental reform'. The CoA would seem to be the appropriate body to initiate the next stage of the process and to define the agenda for achieving the objective.

The agenda for the next round of agricultural talks, the so-called 'mini-round', is already full. First, it will be necessary to review the workings of the Agreement and the progress of transition laid out in the Schedules. Second, the mini-round will have to deal with the remaining anomalies, such as the postponement of tariffication for rice in Japan, Korea, and the Philippines and for some products in Israel. Third, it will have to decide on the next step toward the greater market-orientation promised at Punta del Este. All this will need to be done in addition to any new items that arise.

The strategy for the continuation of the reform process will need to encompass additional market access provisions, further reductions in export subsidies, and more discipline in the area of trade-distorting domestic subsidies. The liberalization of market access itself can presumably use two possible approaches: reduction of tariffs from the high levels at the end of the transition period and increases in the tariff rate quotas specified under the guaranteed access agreements. Domestic subsidies can either be curbed through the further use of AMS constraints or the tightening of the definition of the green box.

Market Access

The level of protection against imports is still high in agricultural markets relative to the trade barriers in manufactures. The question is what process could lead to a removal of this discrepancy in any but the longest time period. How does one get from tariffs of 200 per cent or more to levels of 5–10 per cent, or, more ambitiously, to free trade in agricultural products? It is sobering to see what cuts would be required to achieve this target. A tariff of 200 per cent, not rare for agriculture in the post-UR world, would need a cut of 10 percentage points a year over the 20 years from 2000 to 2020. This looks to be a tall order: the average rate of reduction negotiated in the Uruguay Round for a 200 per cent tariff amounted to 12 percentage points per year, but only for a very limited time period. However, there will not be an easy way around making further significant tariff cuts after the mini-round.

One option for future reduction of trade barriers is to negotiate 'across the board' major tariff reductions, perhaps aiming for a 50 per cent cut in all tariffs over a five-year period. This may prove difficult for those countries with low to medium tariffs, who would object that they would feel the impact of such cuts more heavily. Alternatively, the tariffs could be reduced on a formula basis, with higher tariffs being reduced at a greater rate. The 'Swiss Formula' which was used for tariff reductions in industrial goods in the Tokyo Round would be a candidate (see Chapter 5). This could be a faster and fairer way to the same end, with much of the 'water' being squeezed out of the high tariffs (and the element of 'dirty tariffication' being removed) in the first stage.[9]

Another benefit of using such a formula approach is that it would reduce tariff dispersion among products. The UR agreement provided for a simple unweighted average reduction of 36 per cent, with a minimum cut of 15 per cent for each tariff line. Many countries took advantage of the option of cutting tariffs on sensitive commodities by the minimum and making bigger percentage cuts on items of less domestic sensitivity. As a result, the process of tariff reduction in the Round may have increased the dispersion of tariff levels.[10] In the mini-round, reduction of tariff dispersion may also be best achieved by applying an approach such as the Swiss Formula. As an alternative, a maximum level of tariff could be agreed to which all higher tariffs would have to be reduced over an agreed period.

In the light of the difficulty in reducing high bound tariffs, one may need to consider an additional element in the liberalization of market access. One could for instance eliminate the present gap between bound and actual rates. Bindings could be reduced to no more than the maximum applied tariff in an agreed historical period (say 1993 to 1998, if the negotiations were conducted in 1999).[11] This would lock in agricultural trade reform in a more effective way. The additional discretionary protection that many governments built into the Uruguay Round tariff bindings would in effect be reduced at a stroke.[12]

In addition to further tariff reductions one could expand the guaranteed market access provisions of the Agreement. The tariff-rate quotas (TRQ) are hardly the jewel in the crown of the Agreement. They created a slew of bilateral arrangements which promises to keep agricultural trade an intergovernmental concern for years to come.[13] Moreover, they have created a new wave of governmental interference with agricultural trade through licensing procedures, and provided a playground for rent-seeking traders – who will in turn have an incentive to lobby for the continuation of the high above-quota tariffs. Expanding these TRQ is a simple way of reducing their importance and lessening the impact of the high 'above-quota' tariff. Doubling the minimum access quantities, for instance, would make many of the high tariffs irrelevant. An increase in TRQ, say, of 1 per cent of the level of domestic consumption in each year over a five year period would remove much of their negative effect. In most markets the quotas would become non-binding before the five-year period was over. In effect, tariffication would have taken place at the level of the reduced tariff applicable to the TRQ.

As countries implement the current agreement, the administration of tariff rate quotas (where they are not specific to country of origin) merits particular attention. Issuing quota licences to domestic importers such that they can earn rents is not a very satisfactory method of allocation. However, issuing them to exporting countries in proportion to what they would supply in the absence of trade barriers is impractical. The most convincing approach for handling TRQ would be to auction them on the free market, and to make licences freely tradable once they have been auctioned. Rents would then be skimmed off by the government, and exporting countries could compete on the basis of offer prices (see Chapter 8).

As far as the Special Safeguard Provisions are concerned, agreement should be sought in the Committee on Agriculture that, wherever technically possible, trigger prices should be identical (or equivalent) to the external prices used by the governments concerned in calculating initial tariff equivalents in the UR. Governments had a tendency to use the lowest feasible external prices for calculating tariff equivalents. Hence, using the same prices as trigger prices for the Safeguard Provisions would make sure that additional duties are not used too often, and are not set too high. It will also be important to phase out the use of the Special Safeguard Provisions after the period of transition. This could be done by gradually adjusting the percentages in both the quantity trigger provision and the price trigger provision year by year, so that the safeguards are less and less likely to cut in.

Export Competition

The issue of export subsidies is unlikely to go away. The practice of subsidizing exports of agricultural products has been allowed to continue in a reduced form, but market disruption will still pose problems. Countries that import agricultural products have been the gainers from the subsidies, but even among these countries the disturbance of the domestic market has caused problems. In the next round of negotiations, it will be more difficult than ever to persuade countries that export agricultural goods with little or no subsidy to allow others such as the EU and the US to continue their market-distorting practices. A further push to rein in these subsidies is likely to be high on the agenda.

In addition to the overt export subsidies identified in the Uruguay Round, agricultural exports are often assisted by export credits and credit guarantees. These are clearly forms of export subsidy, given to export firms by governments trying to expand trade. The practice was criticized in the Uruguay Round, and as part of the UR Agreement an understanding was reached that governments should 'work towards the development of internationally agreed disciplines'. Discussions on the subject of export credits in general have been continuing in the OECD, and it is possible that agricultural export credits will eventually be brought into conformity with those in other areas of trade.

The prerequisites for dispensing with export subsidies are a renewed confidence in world markets, with firmer and more stable price levels for the major products, and reduced dependence on intervention-buying in domestic policies. The former condition depends on the success of the Agreement in increasing trade and reducing protection. As for domestic programmes, it is possible that practice and sentiment in both the US and the EU may have moved from the use of market support policies to other instruments by the turn of the century. If that were the case it could be politically easier to get effective curbs on the use of export subsidies by the time of the mini-round. A new set of negotiations could, say, set the target to phase out export subsidies over a ten-year period by 2010. If successful, export subsidies for agricultural products would have been relegated to an inglorious place in trade policy history.

Domestic Support

Domestic support poses a somewhat different problem. Most developed countries have modified their domestic agricultural programmes in recent years to improve the targeting and reduce the output-increasing nature of farm income supports. Developing countries, by contrast, still use output-enhancing instruments because they wish to increase output. At present the rules appear to be written with developed countries in mind. The challenge for the mini-round is to strengthen the rules for domestic subsidies so as to place further curbs on developed countries' output-enhancing producer subsidies, while at the same time allowing developing countries to give adequate incentives to develop their agricultural potential.

The 'blue box' containing the US and EU direct payments that were granted exemption from challenges under the Agreement poses a further challenge. Many countries regarded this arrangement as a bilateral

deal to avoid the task of modifying domestic programmes. The policies themselves are under review for internal reasons. The task for the
mini-round will be to try to ensure that the EU and US modify their
payments such that they meet the conditions laid down in the green
box. In addition, the concept of the green box itself may need to be
refined to give clearer guidelines for policy development.

Regional and Supra-Regional Trade Pacts

At the low point of the Uruguay Round, in December 1990, when the
'final' negotiating session in Brussels had collapsed in disarray, the
negotiation of regional trade agreements (RTA) seemed to be reaching
a peak of activity. The US and Canada were negotiating an ambitious
free-trade area with Mexico to expand the US–Canada Free Trade
Agreement which had just come into effect. The North American Free
Trade Agreement (NAFTA) itself was seen as a part of a wider strategy for the whole of Latin America involving debt and investment
issues as well as trade. The Enterprise for the Americas Initiative (EAI)
was to tie the countries of the continent into a series of trade and
investment agreements, leading eventually to a Western Hemisphere
Free Trade Area to rival the European trade blocs. In Europe, the countries
of the European Free Trade Association (EFTA) were negotiating the
formation of the European Economic Area (EEA) with the EU, to encompass all 17 countries in the largest and richest internal market for
goods and services in the world.

These developments spawned a growing apprehension about the prospect of a weakening of the multilateral trading system. The emergence
of a large number of regional trade blocs led to concern that they
would then be in a position to develop their own trade rules. The fear
was that these blocs would develop protectionist tendencies toward
each other, and thus force independent countries to take shelter within
a regional bloc. The benefits from a broader multilateral trade system
could over time be reduced, even if trade stayed 'open' within the
blocs. The growth of regionalism and intra-regional trade has indeed
been evident in many parts of the world, but the drift towards a world
of trading blocs in recent years has not as yet posed any significant
challenge to the GATT/WTO system (WTO, 1995b).

Agricultural trade is influenced in different ways by the growth of
regionalism. Many regional trade agreements have left agriculture out
of the free-trade provisions, in deference to the political sensitivity of
the sector and the potential conflict with domestic policy objectives.

However some RTA have found that the significance of agricultural trade to one or more parties was too much for the sector to be ignored. The most prominent example of this was the European Economic Community which in the 1960s developed the Common Agricultural Policy on the principle of free internal trade in agricultural products, albeit subject to high levels of protection from outside supplies. The result was to generate significant trade diversion as high-cost European production replaced lower-cost imports from other parts of the world. This illustration of the dangers of trade preferences in RTA serves as a reminder that partial free trade can have untoward consequences. Fortunately, no other RTA have come close to replicating the structure and protection levels of the CAP.

The issue of the treatment of agriculture in a world of trading blocs is likely to become more important. In part this is because the blocs themselves may begin to assume a role in the negotiation of agricultural rules. There are reasons to believe that such blocs will tend to adopt uniform policies toward third countries even if they remain free-trade areas rather than customs unions. But even if the trade blocs remain subsidiary to the multilateral trade system, the treatment of agriculture and agricultural policy within the blocs is of interest since so much of the world's trade will be subject to their rules. These rules can in turn lead to uniform bloc policies, or coordinated national policies and national treatment for partner supplies. The members of the free-trade area could also change to policies that rely on simple border tariffs and decoupled payments, as a way of avoiding intra-bloc conflicts. In the post-UR world, this last option is entirely consistent with the multilateral system.[14]

In this connection, it is interesting to consider the linkages between the GATT approach and these internal policy choices within regional schemes. Exclusion of agriculture from RTA is not a long-lasting solution: it threatens to create the same conflicts within the trade pacts as it has in the GATT. Movement toward common bloc agricultural policies is a possibility, which reinforces the need to have a strong GATT to referee the development of such policies if conflicts arise between the blocs. Full integration of agriculture in RTA is easier the less the policies of the participants rely on trade-distorting instruments. The movement toward decoupled payments will now be made much more attractive by the UR Agreement, as the move to 'green box' payments would simplify the negotiation of RTA.[15]

The most important link between RTA and world agricultural markets may be through the impact of freer regional trade on the reform of national domestic policies. Countries will be under additional pressure to modify these policies so as not to cause tensions among regional trading partners. The fact that such policies are now constrained as a result of the GATT Uruguay Round may make it easier to deal with them in regional agreements. The Agreement on Agriculture could well facilitate the full inclusion of agriculture in regional trade pacts and speed up the reinstrumentation of such policies away from the price guarantees which interfere with trade.

What are the prospects for further trade liberalization in agricultural goods through the medium of regional trade agreements? There are reasons to think that this is a potentially fruitful avenue. The precondition for progress is that agriculture be included more fully in free-trade agreements than previously. This seems to have been the case in recent agreements such as the NAFTA and MERCOSUR.[16] The change has come about largely by virtue of the altered commercial interests of the countries concerned. The major exporters of agricultural products used to identify their export interests with intercontinental trade flows and tended not to look aggressively for regional markets. Only recently have many agricultural exporters come to recognize the importance of regional markets for exports of farm products.[17]

How does this new-found willingness to include agriculture in free-trade arrangements assist in trade liberalization? There is both a direct and an indirect link. The obvious direct link is that trade between the members is liberalized. By the year 2005 there will be virtually no agricultural trade barriers left between the US and Mexico. Canada could catch up with that schedule for the high-tariff commodities by means of reductions in tariffs on intra-NAFTA trade. Other countries may well join the NAFTA by early in the next century. If several of the countries of Central Europe become members of the EU around the turn of the decade, and the same accession arrangements as were negotiated with the recent EFTA countries are followed, there will be an agricultural free-trade zone of perhaps 21 or more countries in Europe by the year 2005. Trade in agricultural products will flow relatively freely throughout most of Latin America, Central America, and the Caribbean if the somewhat ambitious timetables set under the various regional trade initiatives are followed. Asian trade liberalization in agricultural products may lag behind, but could be also underway within the same time frame. Considerable trade creation, and also some trade diversion, can be expected from these regional schemes.

The indirect link between regional free trade and multilateral liberalization is through the impact of the partial opening of borders on national agricultural programmes. This link operates at four different levels. First, there will be pressure to move to policy instruments that do not distort competition within RTA. This suggests that each regional trade agreement will develop in effect its own green box of policies which are deemed to be internally acceptable. Second, there will be pressure to harmonize internal regulations and standards as a way of reducing internal transactions costs and reducing the potential for conflict. Third, trade policies, even where nominally decided at the national level, will have a tendency to fall into line. Rules of origin are likely to be difficult to enforce with homogeneous commodities, and even if they were to be effective a certain amount of trade deflection could still occur.[18] This tendency for external policies to converge even in free-trade areas is reinforced by the likely development of coordinated policies toward third-country trading partners in negotiations. Fourth, there is an inevitable weakening of national policy instruments through the arbitrage possibilities of regional free trade. Free trade will make it more costly to run national farm programmes that involve export subsidies and supply controls, and that try to stabilize the domestic market. This form of 'neo-functionalism' through arbitrage may in the longer run be the most significant driving force behind the change in national policies.

One other development of major significance to the trade system has been the attempt to set up trade agreements which reach beyond the existing free-trade areas. The three most prominent examples of this phenomenon are the Asia-Pacific Economic Cooperation (APEC), the Free-Trade Area for the Americas (FTAA) agreed at the Summit of the Americas in Miami in December 1994, and the recently proposed Trans-Atlantic Economic Area of which one component might be a free-trade agreement (TAFTA). This new breed of supra-regional agreements is different from more conventional regional trade blocs in that they join rather than isolate continents. The fact that a country can be a member of more than one such agreement is the key in this regard. On the other hand, they appear to be more than just *ad hoc* inter-bloc agreements: they include individual countries as well as existing trade agreements as components, and deal with more than just inter-bloc matters. They pose both a challenge and an opportunity for the trading system.[19]

It is this new type of supra-regional trade architecture that is particularly interesting from the perspective of international trade in agricultural goods. It is unlikely that a three-bloc world of Europe, the Americas, and Asia would deal very effectively with the issues of world trade in agricultural products. The blocs would be bound to take very different views on such trade, and inter-bloc tensions would prevail. The historical tensions would all be preserved at the interface between the continental blocs. By contrast, the supra-regional blocs will not be able to dodge the issue of agriculture: it will be internalized within the blocs. The US would be a member of three trade pacts which together would cover the major markets for its agricultural goods. The pressure to include agriculture will come from the lack of political support for any agreement that excluded such an important sector of US exports.

The pace of liberalization of agricultural trade may in fact be set by these supra-regional agreements. If the timetables announced with a fanfare in 1994 are followed, agricultural trade within the Americas and within the APEC will be free long before the multilateral process will have been able to negotiate down the high post-UR agricultural tariffs. Europe is unlikely to be granted less favoured access to US markets than the Asian countries, and would have to open up its own markets to ensure such access. Most present intercontinental agricultural trade flows will be internalized within these supra-regional pacts. Moreover, the pacts themselves are likely to develop of necessity similar modalities for dealing with different standards and domestic policies. The difficulty of having separate trade regimes for each of the supra-markets would soon make for consistency in trade policy.

3 NEW ISSUES FACING THE TRADE SYSTEM

Dramatic changes in world trade have come about as a result of a set of loosely related developments in the late 1980s. These developments include the end of the Cold War, the adoption by many countries of more market-oriented trade systems, the changed view of the role of governments in developed countries, the increased willingness of countries to group regionally to encourage trade and investment, and the increasing concern about environmental costs and the sustainability of economic growth. From these changes has emerged a different international trade structure from that of the postwar period.

Until recently, international trade was influenced by the realities of global political and strategic interests. The GATT was written by the Western powers in the postwar period and gradually extended to other states in the non-Communist orbit. Commercial relations between countries reflected to a large extent the Cold War balance of power. The end of the Cold War has opened up a new set of economic relationships among countries which traded little and has weakened the linkages among others. Trade between the developed market economies and those of Central and Eastern Europe, once heavily restricted in large part to avoid the transfer of technology, is now encouraged. Preferential trade regimes associated with former colonies and political allegiances gradually lose significance as trade is liberalized at the global and regional level. The GATT/WTO has expanded its membership and is about to include China, with the possibility that Russia may be a member before long.

This new structure will pose opportunities and challenges. Among the opportunities is the possibility of creating a truly global economy embracing all the world's significant economies. This in itself is likely to provide a continuing boost to economic growth. The challenges include the more intense relations between countries of different income levels, social systems, and political practices. Trade can easily become entangled in the web of political relationships and it may be tempting to employ trade restrictions as an instrument of diplomacy. This is likely to be particularly true in areas where domestic constituents have global agendas, and see trade as an aspect of such an agenda. The nature of trade in agricultural goods is itself changing in a way that may exacerbate some of these conflicts. This chapter ends with a discussion of the changing nature of agricultural trade and of the 'new agenda' of health, safety, environmental, and labour standards, and the issue of competition and state trading.

The Changing Nature of Agricultural Trade

The nature of trade in agricultural goods is changing over time, in a way that tends to expand its importance in the multilateral trade system. These changes will tend to influence both the political balance of interests in the formulation of trade policy and the interests of individual countries in the outcome. Until recently it has been possible to view agricultural products as among the small group of 'location based' tradable goods, dependent on soil and climate, though somewhat less location specific than minerals. The process of farming was thought to exhibit relatively low economies of scale; 'sourcing out' was rare (except in such cases as live cattle moving across a border for fattening); regulatory conditions varied, but not usually by enough to offset natural and climatic factors; and innovation rents (for being the first to market with a new product) were rare. Trade among countries in the main homogeneous ('bulk') agricultural products, such as cereals, milk, meat, and sugar, is still based in large part on different natural advantages, although with a heavy overlay of distortions arising from agricultural policies. These bulk products have not generally attracted international investment.

Trade in differentiated products such as processed foods, including fruits and vegetables, is now becoming much more important to world trade, in both developed and developing countries. Unlike bulk commodities, producers of these goods can develop comparative advantages by investment from abroad; they can build markets through quality and name recognition; and the overlay of government price support policy is less important in this trade, and is often against the interests of processors who have to deal with high-priced raw materials. Government regulations regarding quality and food safety standards are more significant. Growing mobility of technology, management, and marketing skills is making it possible to 'acquire' an agricultural and food sector through inward investment, together with the processing activities which transform the product for world markets. Foreign processing firms are searching for reliable sources of raw material at a low cost, and are increasingly looking abroad. Consumer tastes, developed by contact with other cultures or brought by immigrants, have expanded markets for commodities previously considered to be non-tradable. Non-food use of agricultural materials has expanded into new areas with the concern about the environment. All these developments point to the globalization of agricultural trade and the reduction in the importance of natural endowments in determining comparative advantage.

Health, Environmental, and Labour Standards and Agricultural Trade

The complex of issues surrounding health, environmental, labour, and safety standards and trade has received considerable attention in recent years. Conflicts between environmental goals and the trade system emerged in the context of the NAFTA and Uruguay Round debates, in particular in the US. Environmentalists worried that trade policy was inimical to environmental concerns, and traders were concerned that environmental policies would inhibit trade flows. As a result of this debate, it was agreed that the GATT/WTO should set up a structure to examine these issues. As the ancient causes of trade conflicts recede, agricultural and food trade will inevitably be 'caught up' in these new issues. The growing proportion of such trade, which involves differentiated quality products and those with high value added, increases the potential for clashes.

Health and Safety Standards

Agricultural trade, unlike that of many other products, has always contained the possibility of the spread of pathogens. Crop seeds, fruits and vegetables, and all livestock products need to be inspected for disease-carrying organisms, as well as for chemical residues and product contamination. A reliable regulatory framework is a prerequisite for engaging in trade in these products. Recently the trend has been towards a more publicly visible process for ensuring plant, animal, and human health and safety. Regulatory systems are scrutinized by consumers, including those abroad. This is in part a function of growing awareness of the health risk of practices and material previously taken as safe. In part it reflects also a scepticism on the part of consumers about the willingness of the industry to regulate itself effectively. In addition, consumers are willing to pay a premium for goods with particular characteristics deemed to be healthy or otherwise beneficial. The result has been tension between consumer watchdog groups, food companies, and regulatory authorities.

The concern with health issues has increased as the nature of trade changes to involve more consumer products. Bulk agricultural goods destined for processing excite little interest among the public, though producers have always been concerned about the spread of plant and animal disease. The increased trade in foodstuffs brings the health problem closer to the consumer. The advent of biotechnology as a major force in agricultural production will give increased scope for such conflicts. These issues are therefore likely to continue to grow in importance. Whether they prove to be a source of trade friction between countries depends on the success of the rules, including the new rules contained in the SPS Agreement concluded in the Uruguay Round.

Environmental Standards

Among the many issues which make up the nexus between trade and the environment one can distinguish four principal types of problem which have been posed for the trade system in recent years. The first of these is the fear that international trade itself may have a negative effect on the environment. This issue has more rhetorical than substantive appeal, and can only be addressed empirically. As economists were quick to point out, trade sometimes has benign effects on the environment, as when the output from a polluting industry is replaced by imports from a non-polluting source and global output is produced with less resource input. Sometimes the impact on the environment can be negative, as would be the case if the output from a polluting industry were to increase as a result of foreign demand (Anderson, 1992).[20] Since one cannot say in abstract which way the effect will be, no *a priori* case can be supported that trade hinders (or helps) the attainment of environmental objectives. On a broader level, increased international trade is of course associated with industrial expansion, which in turn uses raw materials and energy. However, it is not clear that interfering with this aspect of the process of industrialization and urbanization leads to a more environmentally friendly outcome. Indeed, a more convincing general argument has been made that trade helps growth, thereby reducing the poverty which leads people to neglect environmental sustainability.[21] Matters like these need to be discussed at the international level, but it is unlikely, and undesirable, that they should have a noticeable effect on future efforts to continue with liberalizing international trade.

A second problem, the impact of the operation of internationally agreed environmental policies on the trade system, is of considerable concern to those administering such programmes. Coordination among international agencies is mandated by the Rio Declaration and by the WTO Agreement. The search is on in these organizations for instruments of policy that can most efficiently achieve environmental aims without disrupting trade patterns unnecessarily. In many cases these will take the form of non-trade measures, either production regulations or product standards for consumer protection. In some cases it may be necessary (or politically convenient) to employ trade measures to achieve internationally agreed objectives, as is already done in the case of the Convention on International Trade in Endangered Species of Wild Fauna and Flora (CITES) and in the Montreal Protocol on Substances that Deplete the Ozone Layer.[22] Similarly, trade sanctions may have a place in the enforcement of international environmental policies, even if those policies did not themselves involve trade instruments. If the international community as a whole were to decide that collective obligations accompany the benefits from the trade system, then withholding the benefits of trade could be seen as a way of enforcing reluctant countries to shoulder their international responsibilities. The issue then becomes one of strategy rather than of principle. However, such strategic use of trade sanctions at the multinational level could have the negative consequence of encouraging their unilateral use in pursuit of domestically defined environmental objectives. What is needed is a clear institutional structure at the international level for taking joint decisions on multilateral trade sanctions. This is a set of issues worthy of efforts in the WTO.

A third issue is the use of trade sanctions by one country to support the effectiveness of its own environmental policies. In this context two types of environmental problems have to be distinguished (Meinheit, 1995). First, where global environmental effects are concerned, trade policies may be useful, if not necessary, to make domestic policies effective.[23] For example, a country that requires its livestock producers to adopt production methods that reduce methane emissions makes a contribution to limiting the global greenhouse problem. At the same time the costs of domestic products rise, and as a result more may be imported from other countries with less strict emission controls. If trade policies cannot be used to flank domestic environmental policies in such cases, the incentives to engage in, and the effectiveness of, domestic environmental policies targeted at international externalities may be greatly reduced. Issues like these have come to the fore as a result of the tuna–dolphin case brought by Mexico against the United States (Esty, 1994). The WTO may be increasingly faced with cases of that nature. The way the tuna–dolphin case was treated by the two successive panels that considered it shows that things keep changing in the GATT/WTO, and that they may change even more in future. The first panel felt that one country's trade measures directed at the environmental effects of another countries' economic activities were not GATT-legal. The second tuna–dolphin panel, by contrast, could not see why countries should not also take measures to protect environmental goods in other parts of the world. It did, though, conclude that the trade restriction adopted by the US in the tuna–dolphin case was not necessary to achieve its objective, and therefore not justified in the GATT. These issues need to be clarified in future WTO dealings with the links between environment and trade, and it will not be easy to find convincing solutions. Trade access is sometimes seen politically as a privilege which countries can offer to others on conditions of their own choosing. Under such a regime it is possible that a large country can make others change their policies as a condition for having access to its markets. Such a regime is, however, clearly inconsistent with the international trade system of the GATT/WTO, where all members have access rights that can only be abridged in closely defined circumstances. On the other hand, it is important not to undermine incentives for countries to engage in policies that are friendly to the global environment.

The situation is different when it comes to policies dealing with local environmental issues (such as, for example, soil erosion). Domestic producers often argue that trade policies are needed in these cases too, as environmental regulations increase their production costs and therefore reduce their international competitiveness. At the conceptual level, trade restrictions in such cases are neither necessary to achieve the national environmental objective,[24] nor any more justified than protection against competitive advantages that producers in other countries may have because they benefit from a more favourable climate. However, in practice it is not always clear whether environmental issues are of a local or a trans-border nature, and guidelines may need to be worked out in the WTO on how to classify different cases of environmental problems.

The fourth issue is that of the use of environmental regulations as hidden trade barriers – a common complaint of exporters into developed country markets. This has been a particularly contentious issue in the case of agricultural trade, both in livestock and in fruits and vegetables. Producer groups will naturally try to influence the operation of domestic regulations in a way that benefits the market for domestic commodities. Foreign producers are quick to suspect the motive of protection whenever a regulation acts against their sales. The modifications to Article XX encompassed in the Uruguay Round Agreement on Sanitary and Phytosanitary Regulations (as discussed in Chapter 8) were aimed at making it easier to distinguish between genuine and protectionist trade barriers.

This is an aspect of the more general problem of the operation of the trade system in the presence of disparate national environmental regulations. Producers not only feel that the foreign government is denying them access but that their own government is preventing them from competing. The argument is often phrased as one of 'fair' trade. To compete on a 'level playing-field' is sometimes seen to require similar regulations in all trading countries, and suggestions have been made that governments tax at the border imports from those with 'lower' standards. This notion poses perhaps the most immediate threat to the trade system. Producers naturally see many environmental and other regulatory constraints as a burden. If producers in other countries do not have to adhere to the same strict regulations then the local producers will claim unfair competition, pressuring legislatures to grant them offsetting protection. If they in turn are hampered by another country's regulations then they will claim that they are being unfairly denied access. The government will then be under pressure to open up foreign markets. Thus environmental regulations become entangled in trade discussions.

Without a clear way of handling the conflicts that arise from the existence of different regulations the trade system is at risk. Moreover, the same issues arise in a broader context when considering health and safety regulations, labour standards and working conditions, and animal rights. Whenever countries that trade with each other have differing standards and regulations, the possibility exists of trade friction. The issue then becomes one of either harmonizing regulations so that differences are reduced, mutually recognizing the regulations of other countries (with the possibility of a 'race to the bottom' to reduce the cost and hence the effectiveness of national regulations), or living with the differences in standards and ensuring that they are not used to give preference to domestic industry.

In working at the issues involved in the relationship between the environment and trade, the GATT made a start in 1991 by reviving the Group on Environmental Measures and International Trade, which after the Uruguay Round was turned into the Committee on Trade and Environment. The WTO will have to keep environmental issues high on its agenda, and work towards multilateral guidelines and rules for appropriate policies. The outcome of such work one day may need to be a new agreement on environmental policies and trade to go along with the UR agreements on technical barriers to trade and on sanitary and phytosanitary measures.

Labour Standards and Agriculture

Of all the 'new' items on the agenda of the WTO, the most contentious is likely to be the question of labour standards. The argument still reverberates between those who wish to use trade agreements to reinforce the somewhat weak international constraints on labour practice and those who fear that any move in this direction will precipitate a flood of complaints with protectionist rather than humanitarian intent. Many of the same issues arise as in the discussion of fair trade in connection with health, safety, and environmental standards. To what extent can one country dictate the internal labour standards of another country? How can one distinguish between low labour costs and the impact of lax enforcement of regulations? How does one draw the line between ends and means, when the means of some parties are the ends of others?

Agriculture has not yet been caught up in such arguments about labour standards. With the exception of some consumer activity in support of labour action in the grape sector of California, there is little evidence to date that consumers care about the labour conditions under which agricultural goods are produced. It seems unlikely that this holiday will last. As processed farm products continue to penetrate other markets the conditions of their production will become more of an issue.

The threat to the workings of the trade system is perhaps more real than in the case of different product standards. If one government chooses to impose higher quality standards on imports for scientific or other reasons, then consumers lose to the extent that they do not appreciate the extra quality. If a government puts a restriction on the importation of goods produced in particular ways, then the consumer again has to pay more to buy the imported good. But the likelihood of 'capture' of the regulatory system for the purposes of supporting the income of competing domestic producers is greater in the latter case – hence the dislike for 'production and processing method' regulations in the regulation of trade.

Competition and State Trading

As tariff and non-tariff barriers to trade come down, and as the international flows of goods and services increase, the question arises of the impact on trade of different regulations governing competition in different countries. As with the issues of the environment, this topic also has many facets, and the institutions that will eventually be needed to deal with these matters have yet to be devised. Some commentators see the resolution of the problem of the proliferation of anti-dumping laws, with their protectionist tendencies, to be a multilateral code governing competitive practices. Companies that sold goods in other markets would be held to a standard that prevented predatory pricing, and hence not be able to gain 'unfair' advantage over competitors. Others see scope for the regulation of cross-border firms, which might be able to escape domestic competition policies, though in general the more liberal the trade regime the less scope there is for firms to exert market power. Increased international investment flows may make it difficult for governments to assess the competitive impacts of mergers. The conflicts that can arise between countries with different antitrust laws and business practices are apparent in the US objections to the Japanese structure of inter-firm linkages known as the *keiretsu*. It may take many years for these interrelated issues to be covered by international agreement.

One aspect of the interface between trade and competition regulations is, however, relatively straightforward, and is likely to rise to the top of the agenda. This concerns the tensions that are generated when private traders in one country have to compete with state-sanctioned monopolies in others. One of the major issues in agricultural trade for the next decade is likely to be the behaviour of state trading entities and the extent to which they can be integrated into a trading regime postulated on the premise of trade taking place between private firms.[25] State trading has been common in agricultural markets, both in importing and exporting activities. It is not confined to developing countries or to the centrally planned economies, as the existence of marketing boards in such countries as Canada and Australia show. The widespread privatization of agricultural trade in developing countries, and the change in economic paradigm in the formerly centrally planned countries, will tend to reduce the problem over time. However, where such state trading entities still exist, trade frictions are likely to get worse: it is precisely this shift toward private trade that will make state trading a particular problem. As developed and formerly centrally planned countries become more closely integrated into the multilateral system it will become necessary to develop rules that guide the permissible limits of para-statal trading enterprises.

The Articles of the GATT include reference to the practice of state trading. Article XVII enjoins countries to ensure that their state trading enterprises, in both import and export situations, act in a nondiscriminatory manner and base their purchases or sales solely in accordance with commercial considerations. The problem is twofold. First, the existence of the state trading activity itself often reflects precisely the desire to act in a discriminatory fashion. An import monopoly is presumably set up in order to behave differently from private traders. To force commercial rules on such a firm would conflict with domestic aims. Similarly, an export monopoly is set up to take advantage of its market power. Second, the task of discovering whether the state trader has conformed to the Article is well-nigh impossible. By the nature of such enterprises, the test of commercial behaviour is unlikely to be conclusive.[26]

The solution to this problem is likely to rest in the direction of a meshing of national antitrust legislation, international codes, and the new provisions on anti-dumping, subsidies, and dispute settlement procedures that have now been incorporated into the GATT/WTO.[27]

4   CONCLUSION

The long slow march towards the establishment of a liberal, rules-based trading system for agricultural products has finally reached a point at which progress can be measured. The road led from the early drafts of the postwar trade system rules of Keynes and Meade, which would have included agricultural products, through the concern of the British government that the Empire would be compromised, to the steadfast opposition of the US Congress to the infringement of its right to enact farm legislation, on past the establishment of the variable levy of the EU to the dramatic Reagan proposal for an end to all trade-distorting subsidies, to the CAP reform of 1992 and finally to the signatures on the WTO with its new agricultural rules. The road has been marked by intransigence on the part of domestic politicians and reluctance on the part of agricultural officials to see the sector incorporated in the mainstream of the GATT.

That the situation now looks different is due to some remarkable changes in the economic policies of countries large and small and the persistence of a group of agricultural exporters that did not accept that their markets should remain permanently distorted by measures abandoned in other areas of commerce. Most significant, however, have been the changes in the domestic farm policies of the major industrial countries, essentially renouncing the paradigm of market intervention and price manipulation in favour of direct payments to farmers tied less to output. This change is likely to prove crucial in the new agricultural trade regime. The content of these policies should allow a more harmonious development of the agricultural trade system.

Differences between countries over agricultural trade will doubtless continue, but it is likely that they will be more in the area of health, safety, and environmental standards, and in competition policy and state trading, than in the frictions generated by the clash in the international arena between the external effects of inner-directed national farm policies and their accompanying trade arrangements. To the extent that future problems of trade in farm and food products are wrapped in these 'new' issues – rather than being sharply focused on the dumping of unwanted surpluses on world markets and the erection of high trade barriers to keep out 'cheap' food from abroad – they will be part of a new international trade policy agenda and less of a sector-specific area of conflict in international economic relations.

Indeed the Uruguay Round may prove to mark the end of an era in so far as it marks the last gladiatorial confrontation between the agricultural superpowers on the conditions of international commerce in the products of the agricultural sector. Such an outcome would be welcome. The treatment of agriculture over the past fifty years in the GATT has not been one of the more illustrious chapters in the post-war evolution of trade law or of harmonious international political and economic relations.

Trade policy itself is, of course, only one of the influences that shape international trade in agricultural products. The broader changes in technology in the farm and food sector, the growth in demand from population, dietary improvements, and changes in consumer tastes, the incorporation into the trade system of the vast territories of China, the former USSR, and Eastern Europe, and the full emancipation of the developing countries will all have profound impacts beyond the reaches of the WTO rules. Nevertheless, in the annals of agricultural trade negotiations the Uruguay Round Agreement on Agriculture may prove to be the harbinger of a new period of progress and peace.

Notes

1 THE GATT'S ORIGINS AND EARLY YEARS

1. The preparatory work on postwar economic cooperation is comprehensively analysed in Penrose (1953) and Gardner (1980). Accounts which focus on international trade policy are given in Wilcox (1949), Brown (1950), and Diebold (1952). Briefer historical accounts are provided in the standard texts on the General Agreement on Tariffs and Trade, notably Curzon (1965), Jackson (1969), Kock (1969), Dam (1970), and Hudec (1975).
2. This illustrous group was led, from the British and United States sides respectively, by John Maynard Keynes and Harry Dexter White. The colleagues whose names are most intimately associated with the development of international commercial policy were James Meade and Harry Hawkins. The political mentor of all these people was Cordell Hull, Franklin D. Roosevelt's long-time Secretary of State.
3. Gardner observed that 'The talks were entirely exploratory – in the spirit of a university seminar rather than of a formal international conference' (Gardner, 1980, p. 104).
4. The 17 countries represented on the Preparatory Committee were Australia, the Belgium–Luxembourg Economic Union, Brazil, Canada, Chile, China, Cuba, Czechoslovakia, France, India, Lebanon, the Netherlands, New Zealand, Norway, South Africa, the United Kingdom and the United States. The Soviet Union was invited but did not participate.
5. The language was later to carry through into the General Agreement. Jackson notes that 'Almost all the clauses in GATT can be traced to one or other of these trade agreements' (Jackson, 1969, p. 37).
6. The official documents are: UN (1946b, 1947a, b, 1948). The text of the Havana Charter is reprinted in Wilcox (1949, pp. 231–327), and an illuminating 'Analysis of Charter Text' is provided in Brown (1950, pp. 391–557).
7. Beyond the differences of principle was also the practical matter of translating agreements into interpretable form. Wilgress observed: 'The questions with which [the delegates] had to deal were so technical that they used an esoteric language unintelligible to the average man or, for that matter, to the majority of the delegates attending the Conference' (Wilgress, 1948, p.30). The work of GATT panels in later years suggests that ambiguities of intent and language were not entirely avoided.
8. This odd term means that every signatory country undertakes to accord to the imports of every other country the most favourable treatment that it accords to any country with respect to tariffs, internal taxes and other regulations of commerce.
9. The addition of Part IV to the Agreement in 1964 can properly be regarded as a decision to spell out the conditions under which trade policy

244

Notes 245

can be harnessed to economic development – a subject which was central and contentious in the drafting of the Charter for the ITO, and which had lain unresolved since the early postwar years.

10. For an extended discussion see Brown (1950, pp. 163–72).

11. Wilcox, one of the US negotiators of the Havana Charter, summarized the reasons for the antipathy towards quantitative trade restrictions as follows: 'Tariffs are the most liberal method that has been devised for the purpose of restricting trade. They are consistent with multilateralism, non-discrimination, and the preservation of private enterprise. . . . Quantitative restrictions, by contrast, impose rigid limits on the volume of trade. They insulate domestic prices and production against the changing requirements of the world economy. They freeze trade into established channels. They are likely to be discriminatory in purpose and effect. [Because] they give the guidance of trade to public officials; they cannot be divorced from politics. They require public allocation of imports and exports among private traders and necessitate increasing regulation of domestic business' (Wilcox, 1949, pp. 81–2).

12. It was recognized also that quantitative restrictions could be sanctioned for a number of other purposes that, at the time, were regarded as being of a more technical character, for instance, to relieve critical shortages of food and other vital products; for the conservation of exhaustible natural resources, including fish stocks and wildlife; to enforce grading and classification systems for commodities; and to implement the provisions of intergovernmental commodity agreements.

13. In fairness, it should be noted that other developed countries had agricultural programmes that they were unwilling to jeopardize by decontrolling trade in agricultural products. At the time, they typically cited balance of payments difficulties as the reason for imposing quantitative restrictions on imports of farm and food products. But there can be no doubt that the ministers of agriculture in a number of European agricultural importing countries must have been glad to observe the US insisting on agricultural import arrangements which would underpin their own domestic farm support programmes.

14. Other US legislation also provided for the use of import restrictions in support of governmental farm programmes. For instance, imports of sugar were limited by quota under the 1937 Sugar Act.

15. In his analysis of the text of the Havana Charter, Brown treats the agricultural exception under the heading 'Quantitative restrictions whose purpose is not protection' (Brown, 1949, p. 464).

16. An obligation to consult was written into the equivalent provision of the Havana Charter (Article 20,3(b)) but was omitted from Article XI of the General Agreement. It is, however, amply provided for by other articles, notably Articles XXII and XXIII.

17. A few minerals produced in isolated communities also met the specified attributes, and they had the additional characteristic that market difficulties could lead to severe local unemployment.

2 EARLY ENCOUNTERS: 1948–60

1. In this period, of the other major agricultural net importers, Britain did not restrict food imports directly, West Germany claimed the balance of payments exemption, and Japan did not become a signatory to the Agreement until 1955.
2. Hudec (1975, ch. 16, pp. 165–84) provides a detailed account and analysis of the interplay during the early 1950s between United States domestic legislation and the response in the GATT to US restrictions on imports of dairy products.
3. *GATT Basic Instruments and Selected Documents* (BISD) Vol. II, 1952, 16.
4. The retaliation was symbolic since flour imports into Holland from the United States continued at much the same level as before.
5. Like the US waiver, the waiver for West Germany was granted under Article XXV:5.
6. The text of the *Draft Agreement on Commodity Arrangements* is in GATT (1955).
7. Patterson (1966, p. 203). Patterson provides an excellent account of the anxieties caused by the early proposals on a common agricultural policy for Europe, pp. 198–212.

3 THE DILLON ROUND

1. The Dillon Round did not generate a distinctive literature. Accounts are typically carried in analyses of the Kennedy Round, to which it was the prelude (see the references cited in Chapter 4).
2. In the years immediately after the war, economic integration was a movement with wide support. European unity was an explicit objective of the Marshall Plan, and the Organization for European Economic Cooperation created study groups to explore the possibilities. In 1947 France and Italy were also negotiating on forming a customs union. What was not known at the time was that during the period covered by the Havana Conference, the United States and Canada were engaged in secret negotiations on the formation of a free-trade area. See Cuff and Granatstein (1977).
3. A Working Party was established within the GATT to examine trade implications of the tariff, quantitative restrictions, agricultural and association provisions of the Treaty of Rome (BISD 6S/70).
4. A lucid account of intra-European and transatlantic political and economic relationships in this period is Camps (1964).
5. Britain persuaded the 'Outer Seven' to form a free-trade area for industrial products. The Stockholm Convention establishing the European Free Trade Association (EFTA) in 1960 was legally questionable under Article XXIV because it excluded trade in agriculture and permitted its members to apply quantitative restrictions in a discriminating manner.
6. A succinct account of the tariff negotiations can be found in Curzon and Curzon (1976, pp. 168–78).
7. The 6 per cent duty on imports into the EEC of one of such feed ingredient, manioc, was not bound until the Kennedy Round.

8. This was perfectly legal under Article XXVIII of the General Agreement.
9. The State Department was in charge of trade negotiations at this time. Thereafter, under the Trade Expansion Act of 1962, Congress placed the conduct of trade negotiations in the Office of the Special Trade Representative, a creation of Congress and answerable to it.
10. The most readily accessible source of the texts of these agreements is Curtis and Vastine (1971, pp. 36–40).
11. The first common prices – for grains – were not set until December 964.

4 THE KENNEDY ROUND

1. There are several excellent accounts of the Kennedy Round. The preeminent works are Preeg (1970) and Evans (1971). These books are written from an American perspective. So too are Patterson (1966) and Curtis and Vastine (1971). A European view is given in Casadio (1973). The place of the trade negotiations in the context of evolving Atlantic and European political and economic relations is analysed in three outstanding books by Camps (1964, 1965, 1966). All these books deal with the agricultural component of the negotiations. The negotiations on agriculture are especially examined in Leddy (1963) and Zaglits (1967). The present chapter draws freely on an earlier analysis by Warley (1976).
2. Schnittker (1970a). John Schnittker was Under-Secretary in the US Department of Agriculture during the Kennedy Round.
3. This would have provided for Atlantic free trade without giving other countries much of a free ride through the most-favoured nation rule. It meant little, however, if Britain was not a member of the Community.
4. Examination of 'appropriate arrangements' for agriculture had already begun in the Cereals Group, established in 1961, after M. Baumgartner had outlined his controversial plan for the organization of world grain markets to GATT ministers. However, discussion was necessarily confined to concepts and principles because the details of the CAP had not been decided. Several years would pass before the Community would table concrete proposals for grains.
5. The relationship of the Baumgartner Plan to the negotiations is discussed in Lewis (1962) and in Warley (1967a).
6. The ideas of the French government on how world commodity markets should be organized were never put forward as a comprehensive and detailed plan. One of the more coherent accounts of the plan is provided by a synthesis of statements made by M. Baumgartner and his colleagues (notably Edouard Pisani, the French Minister of Agriculture) in the GATT, the UN Food and Agricultural Organization and elsewhere, prepared for the first United Nations Conference on Trade and Development. See 't Hooft-Welvars (1965, pp. 459–64).
7. A more coherent explanation of the EEC's negotiating plan than was available at the time was given later by a senior official of the Agricultural Directorate of the Commission on the eve of the Tokyo Round of trade negotiations. See Malve (1972).
8. It would be another three decades before the international community could

agree to place numeric ceilings on agricultural protection. This occurred in the Uruguay Round. By this time the essentials of the concept of the *montant de soutien* had been made operational in the 'producer subsidy equivalent' and the 'aggregate measurement of support', and both would have a useful role in negotiating reductions in agricultural protection.

9. See Curtis and Vastine (1971, pp. 51–65). There was a precedent for assuring access which the exporters thought might be a model of wider applicability. A plurilateral access guarantee for grain imports was secured from the United Kingdom by its major suppliers in 1964. This committed the UK to maintain grain imports from Australia, Argentina, Canada, and the United States at no less than the average of a recent three-year base period, and to take 'effective corrective action' if imports fell below those levels. For a presciently pessimistic analysis of the fragility of this arrangement and the difficulties of using it as a model in the Kennedy Round, see Brunthaver (1965, pp. 51–9).

10. In the early stages of the negotiations, discussion had been framed in terms of a comprehensive arrangement for *grains*. However, the US had only access guarantees in mind for feed grains, whereas for wheat it was prepared to conclude an arrangement that comprehensively covered international prices, support levels, access guarantees, and surplus disposal. By the end of the negotiations, the US was concerned only to renew the about-to-expire International Wheat Agreement with a higher price floor, and to secure a multilateral agreement on the provision of grains as food aid.

11. India had experienced two failures of the monsoon in the decade, and was receiving huge shipments of grain as food aid.

12. Malmgren (1974, pp. 182–97). Malmgren was a former Deputy Trade Representative.

13. Details can be found in US Department of Agriculture (1967).

14. Schnittker (1970b). Howard Worthington, leader of the US Department of Agriculture's negotiating team on agriculture in the Kennedy Round, later said that his government had believed that 'it was neither practical nor necessary to reach into the heart of a country and negotiate changes in its support programme . . .' and that 'it was not necessary for exporters to take on additional obligations.' See Worthington (1969; 1971, pp. 859–71).

15. This crucial point is discussed by Camps (1965, pp. 20–34). Another keen observer writes: 'The Commission's policy in the agricultural part of the negotiations was largely determined by Vice-President Mansholt's determination to find a policy which would improve the chances of getting the common agricultural policy adopted and of making it more difficult to undermine this policy once it was set up.' (Coombes, 1970, p. 202). Of the French interest Coombes writes: 'The French . . . saw the negotiations primarily as a political means of resisting the pretensions of the United States. In so far as they sought any positive results from the negotiations, they were to subject the Americans to greater discipline in agricultural trade and to use the need for a common position on agriculture in the Kennedy Round to force the Germans to accept . . . the common agricultural policy'(p. 177).

5 THE TOKYO ROUND

1. Perhaps because it was only one element in what was called 'the decade of negotiations', the Tokyo Round of multilateral trade negotiations did not generate a large literature. Fortunately, the major work in the field, Winham (1986), is comprehensive and outstanding. Studies that set the negotiations in context and analyse the issues addressed are: Bergsten (1971, 1975); McFadzean *et al.* (1972), and Corbet and Jackson (1974). A slender monograph containing a weighty analysis is Golt (1974). Since the Tokyo Round opened up many trade policy issues but did not solve them, later work is also an integral part of the literature of the Tokyo Round. A collection of papers which comprehensively bridged the Tokyo and Uruguay Rounds is Cline (1983).

2. Trade reform was but part of the larger task of refurbishing institutions to manage economic interdependence in an increasingly interdependent world. In a vast literature, notably perceptive analyses are presented in Camps (1981) and Ostry (1990).

3. A number of the 102 countries that joined in making the Tokyo Declaration were not signatories to the General Agreement. For this reason, the negotiations were technically conducted 'under the auspices of the GATT', and the results of the negotiations had to be formally brought within the GATT's legal framework at the 34th session in November 1979. The term 'multilateral trade negotiations' was intended to indicate that the matters addressed extended far beyond the adjustment of tariffs, which had been the principal focus of earlier GATT Rounds. See McRae and Thomas (1983, pp. 51–84).

4. The comprehensive global strategy and 'Programme of Action' for eliminating hunger and food insecurity is described in United Nations (1975).

5. On a US initiative, the OECD established a high level group of outstanding internationalists, under the chairmanship of Jean Rey, on 'trade and related problems'. In its report (OECD, 1972) the group concluded 'A new effort to secure greater liberalization, secured through negotiation, is needed not only for the direct benefits it will bring but because without it the divisive forces of protectionism will grow stronger, with the risk that we will slip back into an era of restriction and ultimately of contraction of the international economic system' (p. 110). An analysis of the declining effectiveness of the GATT as a global institution in this period is provided in Jackson (1978, pp. 93–106).

6. The first three of these became members of the European Community on 1 January 1973. Following a referendum, Norway decided not to accept the membership it was offered.

7. A former Commissioner for External Affairs of the European Community made the point this way: 'There is an almost absurd disproportion between the expectations of the European Community's partners in the world and the instruments which the Community has at its disposal to respond to these expectations' (Dahrendorf, 1974, p. 65)

8. For a later account of these weighty charges see Patterson (1983, pp. 223–42).

9. Following the formation of the OPEC and the temporary embargo by the

US on exports of soya beans in 1974, Japan did, for a time, have an interest in seeing stronger international disciplines being placed on the use of export restrictions. But this interest soon dissipated and was not pressed in the MTN.

10. Most countries eventually based their tariff offers on an arithmetic formula proposed by the Swiss delegation: $z = 14x \div (14+x)$, where z is the final and x the original tariff. The EEC used the less-demanding constant of 16 in its final offer.

11. In addition to the reports prepared by Roth (1969), Petersen (1971), the Williams Commission (1971) and the OECD (1972), and the books by Bergsten (1971, 1973 and 1975) and McFadzean *et al.* (1972), the pivotal importance of reforming agricultural trade was also a theme in the National Planning Association (1971) and Committee on Economic Development (1971).

12. A representative statement was made by Clifford Hardin, US Secretary of Agriculture: 'We should return to the original promise of the GATT – the promise of a market oriented agricultural trading world. Export subsidies should be eliminated, present protective systems, such as variable levies and quotas, should be replaced by fixed duties, and income objectives should be met by direct payments which will not stimulate production' (Hardin, 1971, p. 800).

13. The interrelations that had developed in the mid-1970s between the negotiations on agricultural trade, world hunger and food insecurity, the new international economic order and East–West relations are examined in Warley (1976, 1978), and Josling (1977b). Earlier studies that had begun to acknowledge the broader agenda include the Atlantic Council (1973) and the Brookings Institution (1973).

14. Winham notes that the importance of the negotiating groups varied. Some of those working on non-tariff measures had before them drafts that had been agreed on an *ad referendum* basis in the work programme begun in 1967. The subgroup on technical barriers to trade was one such. The subgroup on subsidies and countervail duties was the principal forum in which agreement was reached on this complex and difficult subject. By contrast, the Tariff Group was effectively bypassed by the high level political agreement between the US and the EEC (Winham, 1986, pp. 97–127). Negotiations on access and commodity arrangements for dairy products and meat were conducted in their respective subgroups, but much of the substantive negotiations on grains was conducted outside the GATT framework.

15. Congress had repudiated the Kennedy Round's agreement on the US system of valuing certain chemical imports for duty and had gutted the antidumping code by insisting that the Tariff Commission give precedence to domestic legislation.

16. Congress set a high price on the limitation of its authority. Congressional appointees became more directly involved in international negotiations. Industry advisory groups were mandated. Section 301 of the Trade Act provided for unilateral action by the US if other countries' national policies or commercial behaviour damaged US economic interests. Most damaging of all, US trade remedy laws were changed in ways which allowed

the US to develop a system of 'contingent protection' which could readily be used to harass exports to the US. See Destler (1986, pp. 111–22), and de C. Grey (1983, pp. 243–57).

17. 'The rules of the game determine its outcome. Procedure is substance' (Koenig, 1975). Koenig was Agricultural Attaché at the US Mission in Geneva.

18. This was amply demonstrated by the saga of the Flanigan Report. A study prepared by the US Department of Agriculture (Flanigan, 1973), recommended that the US should seek advantage for its grain and oil-seed exports to Europe in part by easing its restrictions on imports of dairy products and meat. This apparently reasonable proposal was greeted by such a storm of protest in Congress and in the country that even the liberal Secretary of Agriculture Earl Butz was forced to distance himself and the Administration from it, and the unfortunate supervisor of Dr Flanigan was reassigned.

19. The text of the what became known as the 'Dent–Soames compromise' is given in Harris (1977, pp. 49–51).

20. The story of this pivotal development is told and documented in Winham (1986, pp. 164–7).

21. These included subsidies and countervail duties, anti-dumping, standards, import licensing, customs valuation, and government procurement. The subjects of quantitative import and export restrictions were not pursued to a conclusion, and negotiations on revised rules for the use of temporary import restrictions failed.

22. The full texts of the MTN agreements are given in GATT (1979, 1980).

23. Australia and New Zealand continued to complain from the sidelines that conditions in world markets for livestock products had not improved. Indeed, they had worsened as dislocations in international markets for grains and oil-seeds spilled over into animal agriculture. The struggle of these countries to keep access issues before the negotiators on agriculture was only partially successful since they carried little negotiating weight.

24. The value of US agricultural exports was $7 billion in 1970 and close to $20 billion by 1974. Such instability was welcomed.

25. It attracted the interest of US Secretary of State Henry Kissinger. In the mid-1970s, the US Department of State played the lead role in the examination of proposals to establish some system of internationally managed grain reserves and to raise the multilateral commitment to providing food aid to the 10 million metric tons per year target set at the 1974 World Food Conference. Secretary Kissinger was also interested in maintaining the access of the USSR to Western grain supplies, and expanding US–USSR grain trade, as part of the policy of détente.

26. A senior official of the US Department of Agriculture (USDA) observed: 'In agriculture . . . the European Community regards the world market as wholly artificial – a place where national shortages and surpluses are cleared with the help of subsidies and other devices paid for by those who can afford it' (Worthington, 1972, pp. 7–9, 16). Of course, the US dairy industry and other import-competing clients of the USDA used exactly the same argument to explain why their protection must be maintained.

27. Changes in trade arrangements for pigmeat and poultry products were handled by requests and offers in the closing stages of the negotiations.

28. World beef markets were in turmoil at the time. In 1974, international beef prices collapsed in response to a cyclical peak in beef supplies and the herd liquidations that followed the surge in feed grain prices. The EEC and Japan unilaterally closed their borders to beef imports. The US and Canada then restricted beef imports to prevent deflected supplies undermining their markets. Canada formally invoked Article XIX, and both justified their actions as being necessary to 'safeguard' domestic industries.
29. The reports of the Williams Commission (1971), the OECD high level group (OECD, 1972), the Atlantic Council (1973), and the Brookings Institution (1973), cited above had all made favourable comments on the approach.
30. The concepts of producer and consumer subsidy equivalents – which were destined to play a role in the Uruguay Round – were first applied to agriculture in 1973. See FAO (1973).
31. The following statements are illuminating: 'it is completely unrealistic to think that the European Community – or the United States for that matter – would make domestic policies the primary focus of negotiations . . . let's start with the border measures and the export subsidies' (Butz, 1974) and, 'The United States will neither seek nor offer commitments on domestic farm price support levels. These are domestic concerns . . . However, the United States could reach agreement on general principles aimed at assuring that domestic policies are not used to offset negotiated commitments on access' (Fraser, 1975, pp. 2–4). Fraser was Assistant Administrator for International Trade Policy, Foreign Agricultural Service, US Department of Agriculture. The position of the Carter Administration was even clearer: 'the United States was not seeking in the MTN to undermine the Common Agricultural Policy (CAP). This recognized the fact that the CAP is an outgrowth of the current economic and political circumstances in Europe and that this basic policy was not susceptible to fundamental change in an international trade negotiation. Our strategy in negotiations with the European Community was to focus our efforts on obtaining access for specific products and on seeking improvements and modification in those international rules where the United States had important interests and where we felt progress could be attained; not to attempt to take the basic CAP head-on' (letter from the US Special Trade Representative, Robert S. Strauss, to Senator Charles L. Vanik, Washington D.C., 2 May 1979).
32. The results of the negotiations on tariffs and quotas are given in the GATT's official records and in national publications. An excellent example of the latter is US Department of Agriculture (1981).
33. The 1971 International Wheat Agreement was due to expire on 30 June 1979.
34. Some of the countries interested in food security, and some large wheat importers, including the USSR, were not participating in the MTN.
35. The negotiations were held under the chairmanship of a Swiss, Arthur Dunkel, a man who would later provide outstanding leadership on agriculture in the Uruguay Round.
36. Hathaway (1983, p. 452). Hathaway was Assistant Secretary, US Department of Agriculture, in the Carter Administration.

37. Canada did no more than agree not to accede to the demand of its milk producers that annual cheese imports be reduced below the then current level of 45 million pounds. This 'concession' was balanced by Canada's choosing not to join the International Dairy Arrangement lest the floor prices for dairy products limit its ability to continue to dump a structural surplus of skim milk powder on world markets. The US increased its cheese import quota by 15 per cent to 110 000 tons. However, since the new quota included cheeses that were not previously subject to quota and whose importation had been expanding rapidly, there was some doubt as to whether exporters' long-term opportunities in the US market had really been improved.

38. The US had proposed that all government 'aids' to industry be categorized according to their production and trade impacts. Such 'generally available' and production- and trade-neutral programmes as aids to research and development, adjustment assistance and staff training would be permitted and non-actionable ('green'). At the other extreme, a category of 'red' subsidies would be prohibited or, if used, subject to countervail and retaliatory action without an injury test. Other ('amber') programmes would be conditionally permissible and actionable only on proof of material injury.

39. The concept of 'injury' was drawn from the GATT Article VI and 'serious prejudice' from Article XVI:1. 'Nullification and impairment' of expected GATT benefits drew on the interpretation of Article XXIII established in previous panel rulings that the GATT should protect the 'reasonable expectations' of the contracting parties.

40. For instance, in grains, the European Community maintained on occasion that its 'restitution' payments were not export subsidies within the meaning of the General Agreement, and the legal status of the United States' grain deficiency payments and Canada's two-price wheat plan remained unclear.

41. US Department of Agriculture (1981, p. 13). The notion of a 'cathedral' agreement on agricultural policies and trade was often associated with the name of EEC Commissioner Finn Olav Gundelach.

42. Originally thought of as a broad updating of the GATT's rules and procedures, the 'framework' agreement ended up as being primarily a vehicle for enshrining the developing countries' special privileges and weaker obligations under the General Agreement. A less controversial provision entailed the clarification of procedures regarding the use of panels in the settlement of disputes (BISD 26S/203–18).

6 MARKETS, POLICIES, AND TRADE RULES IN CRISIS: 1979–86

1. Not all countries shared this view, and major efforts had to be made in order to drag the EC to the negotiating table. The negotiations on the Uruguay Round agenda are described in Chapter 7.

2. Various issues of the European Communities' *Agricultural Situation in the European Community*. The value of the US dollar increased from 0.72 ECU in 1980 to 1.31 ECU to the dollar in 1985.

3. As a first step in 1983, the US had introduced a massive subsidy on wheat flour exports to Egypt, capturing this whole traditional EC market for one year.
4. The domestic political constraint was that strong vested interests had built up around existing programmes, making reform politically costly. The use of external negotiations could help to persuade recalcitrant farm groups that there was no reasonable alternative to initiating reform, and that the fact that other countries were also changing policies lowered the cost. In the negotiations, this domestic intransigence could be used as a bargaining ploy to convince others of one's inability to respond to their demands. However, to play the international 'blame shift' card too often or too blatantly would have the effect of turning domestic groups against the negotiations, which would then limit the flexibility in international talks.
5. Since the MTM report, the estimating and publishing of PSE and CSE values has become routine in the OECD. The method has been refined, policy coverage has improved, and a monitoring and outlook report on *Agricultural Policies, Markets and Trade* has been published annually since 1988. All OECD member countries are now included, except for Mexico for which the data is still in preparation. Updated PSE and CSE estimates are now made available in electronic form to the general public. Meanwhile, the OECD has begun to extend the approach, and its review of national agricultural policies, to the countries of Central and Eastern Europe (OECD, 1994a, 1995).
6. For example, in 1981 the Consultative Group of Eighteen requested the GATT Secretariat to prepare 'an analysis of the rules of the GATT, including the Codes as they applied to agriculture, highlighting any differences in obligations as between agriculture and industry' (GATT, 1982b, p. 1). The resulting document on *Agriculture in the GATT* (GATT, 1982b) was a useful description of the GATT rules on agriculture, reporting also on trade disputes which had related to these rules. However, even though this document did not contain any evaluation of the state of affairs, it was kept strictly confidential.
7. When the GATT was drafted, it was 'intended to apply to trade in agricultural and industrial products alike' (GATT, 1979, p. 18). Also see Curzon (1965, p. 166).
8. The five other GATT provisions where foodstuffs, agricultural/primary products or commodities are mentioned specifically are Articles VI:7 (no material injury caused by domestic price stabilization); XI:2(a) (export restrictions to relieve critical shortages); XI:2(b) (restrictions related to standards for the classification, grading or marketing of commodities in international trade); XX(b) (measures to protect human, animal or plant life and health); and XX(h) (obligations under international commodity agreements). In addition, primary products are mentioned in three places in Part IV of the GATT which deals with trade and development.
9. For an excellent legal analysis of the specific rules for agriculture in the GATT, see Davey (1993). The economic aspects of the difficulties encountered in applying GATT rules for agriculture in dispute settlement are discussed by Hartwig (1992).
10. After the Uruguay Round, the provisions of the General Agreement on

Tariffs and Trade as concluded in 1947, with all subsequent rectifications, amendments and modifications prior to the Final Act of the Uruguay Round, are referred to as 'GATT 1947'. It is legally distinct from the GATT 1994, which in addition to the GATT 1947 includes a number of understandings on GATT provisions agreed in the Uruguay Round and the Schedules of all participating countries. The Uruguay Round Agreement has left all provisions of the GATT 1947 in force, including the agricultural exceptions. However, as explained in Chapters 7 and 8, the new commitments countries have accepted under the Uruguay Round Agreement on Agriculture have effectively superseded these agricultural exceptions.

11. Other domestic measures which may justify import quotas are removal of a temporary surplus (subparagraph (ii)) and supply management of an animal product for which the imported good is a feed (subparagraph (iii)). Subparagraph (ii) has played only a limited and subparagraph (iii) no role at all in GATT disputes. The following discussion will therefore be limited to subparagraph (i), and when Article XI:2(c) is referred to, only that provision is meant.

12. According to a count by Hudec (1993, p. 408), out of a total of 207 GATT disputes between 1948 and 1989, 16 referred to Article XI:2(c).

13. The titles and numbers of GATT complaints given in the following pages correspond to those in Appendix II of Hudec (1993). Hudec provides summary descriptions of the complaints, as well as references to panel reports and related GATT documents. Adopted panel reports are published in the BISD. For a documentation of GATT disputes, see also Pescatore, Davey and Lowenfeld (1991). Rulings of GATT panels, and arguments raised in them, will be reported in the following in a rather summary and nontechnical manner. Any ruling applies strictly to the case concerned only, and rulings of different panels on different, though closely related, disputes have not always been consistent. For the reader interested in accurate information there is no way around working through the sometimes rather hard to read and usually lengthy panel reports. The most important agricultural GATT disputes are summarized and analysed by Davey (1993) and Hartwig (1992).

14. There are two points in the Uruguay Round Agreement on Agriculture which may turn out to have some relation with the 'old' Article XI:2(c). The 'rice clause', allowing certain countries to maintain, for some time, quantitative restrictions for certain commodities, requires that 'effective production restricting measures are applied' to the products concerned. Also, certain direct payments under 'production-limiting programmes' are exempted from the reduction commitments for domestic support, and from the calculation of the aggregate measurement of support. Should disputes arise over these provisions, it is conceivable that panels may look back to relevant elements in past rulings on Article XI:2(c).

15. 'USA v. Canada: Quantitative Restriction on Ice Cream and Yoghurt', Complaint 195, and 'USA v. Japan: Twelve Agricultural Products', Complaint 148.

16. See the disputes referred to in the previous note. For comments on, and sources for, the drafting history of the GATT provisions discussed here, see Chapter 1, and Davey (1993).

17. 'USA v. EEC: Minimum Import Prices', Complaint 76, and 'Chile v. EEC: Licences for Dessert Apples', Complaint 180. However, in the earlier case of 'Chile v. EEC: Restriction – Apples from Chile', Complaint 89, the panel somewhat surprisingly felt that a policy of this type could be considered to restrict domestic supply.
18. 'Chile v. EEC: Restriction – Apples from Chile', Complaint 89.
19. 'USA v. EEC: Minimum Import Prices', Complaint 76.
20. 'USA v. Canada: Quantitative Restriction on Ice Cream and Yoghurt', Complaint 195.
21. 'USA v. Japan: Twelve Agricultural Products', Complaint 148. As Davey (1993) rightly notes, this particular ruling was potentially pernicious as it might have encouraged countries to restrict the imports of more rather than fewer products.
22. Though they have rather different economic implications, these two dimensions have not been very clearly distinguished in the General Agreement, and have not always been treated as separate by panels. In the case of 'USA v. Canada: Quantitative Restriction on Ice Cream and Yoghurt', Complaint 195, for example, essentially no distinction is made between 'like product' and 'in any form'.
23. This is implicitly acknowledged in the definition of the term 'in any form' which, as stated in an interpretative note to Article XI.2(c), 'covers the same product when in an early stage of processing . . .'. The note goes on to require that the product must also be 'still perishable'. Doubts can be raised regarding both the economics behind that requirement and the exact definition of perishability. The panel on 'USA v. Canada: Quantitative Restriction on Ice Cream and Yoghurt', Complaint 195, noting the questionable practicability of the concept, therefore rightly dropped it from its considerations.
24. The debate about the appropriate treatment of processed foods in this context, and in the context of Article XVI:3, touched implicitly upon one of the fundamental problems caused by agricultural protectionism – the potential distortions created in the processing industry. Because of the secular growth of relative importance of the processing industry and of trade in manufactured food products, this factor may well be one of the nails in the coffin of agricultural protectionism.
25. 'USA v. Japan: Twelve Agricultural Products', Complaint 148. This panel finding applies strictly to the dispute concerned only, but it would appear to be a valid observation for most cases of domestic supply control. A similar observation has been made by the panel dealing with 'USA v. Canada: Quantitative Restriction on Ice Cream and Yoghurt', Complaint 195.
26. See the summary description of the complaints in Hudec (1993). Out of the 16 complaints in which Article XI:2(c) played a role, inconsistency with at least one of the requirements in this provision was found in ten cases. In five cases no ruling was made because the dispute had been settled bilaterally, always with the result that the quantitative import restriction was eventually lifted. In one case, the working party dealing with the complaint was unable to determine whether the quota volume had been set correctly.

27. For some quantitative evidence, see the following subsection, Outcome of the Committee's Work.
28. For an analysis of the relationships between GATT rules and this agreement, see Hartwig and Tangermann (1987). For a discussion of voluntary export restraints in agriculture, with an emphasis on the EC–Thailand agreement, see Winterling (1986).
29. As Hudec (1993, p. 23) observes, 'the consensual action of the exporter left no basis for GATT legal action against the importing country'.
30. Davey (1993, p. 6) who makes this point also notes that 'any attempt to apply [this argument] in practice would be difficult. It would probably be impossible to determine what tariff levels would be truly prohibitive for any given product. In the extreme cases, though, the argument has facial appeal.' Later, however, Davey (1993, p. 56) observes that if there is no tariff binding on a product, then a prohibitive tariff can be charged. This . . . is the usual result under the EC CAP's variable levy system.' In other words, even though economically a prohibitive tariff may be equivalent to a (prohibited) zero import quota, it would be legally difficult to prevent a country from setting an unbound tariff as high as it likes.
31. See, however, Davey (1993, p. 6) who argues that 'in some cases the EC's use of variable levies violates the GATT Article X's requirement that tariff schedules be published so that traders know the conditions under which they can trade. For some products (especially wheat), the variable levy is changed almost daily, which certainly goes against the spirit if not the letter of Article X.'
32. 'Uruguay v. 15 Developed Countries: Recourse to Article XXIII', Complaint 54. This rather general lawsuit was 'meant to give dramatic demonstration of how GATT was failing to serve developing countries, due to developed country protection' in the area of temperate zone agricultural products (Hudec, 1993, p. 446). During the panel proceedings, Uruguay requested a ruling on the GATT conformity of EC variable levies and the CAP in general, without advancing any specific legal argument.
33. Based on Hudec's (1993) list, a total of 16 GATT disputes dealt with the exception of agriculture from the ban on export subsidies, this number being exactly equal to the number of disputes dealing with Article XI:2(c). Out of the 16 cases relating to export subsidies, nine addressed Article XVI:3, and six related to Article 10 of the Subsidies Code. In addition there was the case of 'US/EEC: Subsidies on Pasta Products', Complaint 105, which did not address Article XVI:3 because the issue was whether the subsidies concerned were subsidies on a non-primary product prohibited under Article XVI:4. In economic terms, though, this case essentially dealt with a subsidy on the primary product content of a processed product. In addition to complaints addressing the agricultural exception from the ban on export subsidies, there were a few disputes involving agricultural export subsidies which did not question the legality of the subsidy, but related to countervailing duties or to nonviolation nullification and impairment.
34. However, in 1960 a GATT working party developed an illustrative list of export subsidies which was adopted by the CONTRACTING PARTIES. See BISD 9S/185–8. This list was later amended and annexed to the Tokyo

258   *Notes*

Round Subsidies Code. Some elements of this list reappeared, together with new elements, in the definition of export subsidies in the Uruguay Round Agreement on Agriculture (Article 9:1).
35. 'Australia v. France: Assistance to Exports of Wheat', Complaint 50.
36. These three cases were the two sugar cases mentioned in note 37 and 'USA v. EEC: Subsidies on Wheat Flour', Complaint 103.
37. The two sugar cases were initiated in 1978, before the Tokyo Round had been concluded, and were therefore pursued under Article XVI:3 rather than under Article 10 of the Subsidies Code. However, the drafting of the Subsidies Code had been essentially completed at the time, and the panel reports actually made use of the language in its Article 10. See Hudec (1993, pp. 130–4).
38. 'USA v. EEC: Subsidies on Wheat Flour', Complaint 103. The citation is from GATT (1983).
39. Adoption of the panel report was blocked by the EC in the Subsidies Code Committee. The dispute was never formally settled. See Hudec (1993, pp. 147–51 and 491–2), for a description of the dispute and its consequences.
40. In this context it is interesting to note that the GATT working party responsible for drafting the 1955 amendment suggested 'that in determining what are equitable shares of world trade the CONTRACTING PARTIES should not lose sight of . . . the fact that export subsidies in existence during the selected representative period may have influenced the share of the trade obtained by the various exporting countries'. (BISD 3S/226).
41. 'Australia v. France: Assistance to Exports of Wheat', Complaint 50.
42. Hudec (1993, p. 149) adds the nice anecdote that 'off the record, there were lots more bizarre defenses circulating in the corridors of the GATT Secretariat: a high Secretariat official once shared with the author whispers that some shipments invoiced as U.S. wheat flour exports had really been C.I.A. arms shipments hidden in flour bags.'
43. This view would be supported by the following language in the report of the GATT working party having drafted the 1955 amendment of Article XVI: 'the CONTRACTING PARTIES should not lose sight of . . . the desirability of facilitating the satisfaction of world requirements of the commodity concerned in the most effective and economic manner' (BISD 3S/226).
44. This interpretation could be suggested by reference to 'the three most recent calendar years' in Article 10:2(c) of the Subsidies Code, though it is not quite clear whether the requirement that these should be years 'in which normal market conditions existed' does exclude a reference period in which exports were subsidized. Were past export subsidies 'normal', while current export subsidies are not?
45. In Article 10:3 of the Subsidies Code, signatories agreed 'not to grant export subsidies on exports of certain primary products to a particular market in a manner which results in prices materially below those of other suppliers to the same market'.
46. 'US v. EEC: Subsidies on Pasta Products', Complaint 105.
47. See the description by Hudec (1993, pp. 151–4), who characterizes the legal issue in this case as 'the type of problem that law professors make up for examinations'.
48. This was exactly the point made by the earlier US reservation, and eventually

it was implicitly accepted by the US when the pasta dispute was settled with US acceptance of EC subsidization, as long as the EC did not over-compensate pasta producers for the difference between domestic and world market prices of their raw material.

49. Indeed, as mentioned in Chapter 5 the USA had promised to the EC, in a letter from US Special Trade Representative Robert S. Strauss to EC Commission Vice President Gundelach, that 'it in no way intend[ed] to undermine the CAP' through use of the Subsidies Code.

50. There have been other complaints over agricultural export subsidies after the wheat flour dispute (see Complaints 126, 134, 190), but none of them was pursued even to the point where a panel ruling was made. In addition to the disputes mentioned so far, there have been three disputes over countervailing duties against agricultural export subsidies (Complaints 137, 149, and 151) and one complaint regarding nonviolation nullification and impairment caused by an agricultural export subsidy (Complaint 135). As the legality of export subsidies was not at issue in these cases, they are not discussed here.

51. As noted above, in GATT disputes over Article XI:2(c), the defendant country has never won the case. When it came to export subsidies, the situation was nearly reversed. Davey (1993, p. 39) notes that 'in the typical GATT case, the complaining party wins. But not in export subsidy cases'.

52. There were two GATT disputes between the US and Canada over countervailing duties against domestic agricultural subsidies (Complaints 203 and 205). These disputes are not discussed here either.

53. 'Australia v. EC: Production Subsidies – Canned Fruit', Complaint 101, and 'USA v. EEC: Production Aids – Canned Fruit', Complaint 107.

54. A panel ruling was made only in the US–EC dispute. Australia did not further pursue its complaint when the US–EC dispute was settled through a commitment by the EC to reduce the level and scope of the subsidy. See Hudec (1993, p. 489).

55. The original AG/DOC series was an inventory of non-tariff measures es-tablished in 1975, following an earlier GATT inventory of non-tariff measures created after the failed attempt of disciplining the use of NTM during the Kennedy Round.

56. In a sample of notifications by thirteen major countries (ten devel-oped, three developing), there were 1529 measures affecting imports un-der the headings of 'levies and other charges' and 'licensing and import restrictions'. Of these 1529 measures, only 28 were classified as 'result-ing from the lack of observance or application of certain [GATT] provi-sions'. 177 measures were classified as being covered by waivers or grandfather clauses. Only 19 measures (of which 16 were maintained by Canada) were classified as being taken by virtue of Article XI:2 of the GATT.

57. The first version of the Draft Elaboration was produced in 1984 (GATT, 1984). The paper was later revised three times, and the final version was issued in 1986 (GATT, 1986). Annexed to that document was a 'non-paper' drafted by the CTA chairman.

7 THE URUGUAY ROUND NEGOTIATIONS

1. The setting for trade policy in the lead-up to the Uruguay Round is well documented in Cline (1983), Hufbauer and Schott (1985), Tumlir (1985), Eminent Persons Group on World Trade (1990). For an overall assessment of the Uruguay Round, see May (1994), OECD (1994b), Schott (1994), Senti (1994), Croome (1995), Hauser and Schanz (1995). An overview of the agricultural negotiations can be found in Ingersent, Rayner and Hine (1994). Literature on the results of the agricultural negotiations of the Uruguay Round is referenced in Chapter 8.

2. Not everyone expected the talks to be over in four years. An article in the *Economist* (13 September 1986, p. 63) speculated that 'the trade talks that begin in Uruguay on September 15 could last a decade'. This pessimism was reinforced by the notion that a decision 'to include agriculture and trade in services [could] by itself cause the talks to founder'.

3. The Cairns Group members presented a cosmopolitan front to the major powers, united in their dislike of the impact on the world market of US and EC agricultural policies. Members of the Cairns Group were (and are) Argentina, Australia, Brazil, Canada, Chile, Colombia, Hungary, Indonesia, Malaysia, New Zealand, the Philippines, Thailand and Uruguay. Fiji was a member of the Cairns Group, but not a Contracting Party to the GATT. It therefore took no part in formal negotiations. Australia took the lead in much of the Cairns Group activities, and the presence of Argentina, Brazil and other Latin American countries added to the political weight of the group. Canada was initially an enthusiastic member of the group but disagreed on several key issues with other members, and lost influence in the deliberations. Hungary was on the list, at a time when that country was still in the Comecon system. But Hungary was both an agricultural exporter to countries beyond the Soviet bloc, and was attempting to distance itself from the rest of the bloc in economic policy matters. Cairns Group membership thus helped Hungary and also added to the cosmopolitan aspect of the group. The origin and role of the Cairns group in the negotiations is described below. For a discussion of the formation of the group by an active participant see Oxley (1990).

4. The original political impetus for a new round is generally credited to William E. Brock, US Special Trade Representative under the first Reagan Administration, 1981–4. Academic and think-tank analyses of the troubles of the trade system in the early 1980s include Cline (1983) and Scott (1984).

5. Though of importance to countries other than the US, the interest of that country was necessary to put these issues on the international agenda.

6. Important protagonists for a new round of negotiations included James Robinson of American Express and John Reed of Citybank, who saw the advantages of extending to service trade the rules and dispute settlement procedures that obtained in goods trade under the GATT. Among those particularly concerned with intellectual property rules were the pharmaceutical, software, and film recording industries. Many other segments of US industry, particularly those hit hard by import competition from Asia, were less enthusiastic over further trade liberalization, and some, such as

textiles, had always been against any move to free up trade. The coalition of high-technology, service and agricultural sectors was crucial for the continued support of a reasonably liberal US position in the Round.

7. For an account of this agreement see Rosen (1989).

8. The Agreement was signed in January 1988, and entered into force in January 1989. For a discussion of the negotiations leading to this agreement see Wonnacott (1987).

9. The Multifibre Arrangement (MFA) and its predecessor the Long Term Textile Agreement (LTA) were an attempt to allow a controlled expansion of trade in textiles by the use of quotas on imports into developed markets. Unfortunately, the quotas expanded relatively slowly, and became instead a device for protecting these markets from cheaper textiles from developing countries.

10. Japan also suffered from the EC's contingent protection measures and voluntary export restraints. Like Europe, Japan was not keen on the prominence given by the US and others to agriculture in the new agenda.

11. The EC Commission tabled its White Paper on the Completion of the Internal Market in June 1985. The '1992' Programme, as it became known, included freeing up the internal EC market in services, as well as removing non-tariff barriers within the internal market for goods.

12. The meeting was held in the EFTA building in Geneva, rather than in the GATT headquarters, thus emphasizing its independence from the established GATT procedures (Oxley, 1990, p.140).

13. This compromise was at the time referred to as the 'cafe-au-lait' scheme, in recognition of the late night efforts of Felipe Jaramillo from Colombia and Pierre Louis Girard from Switzerland.

14. The phenomenon of countries meeting in informal groups which crossed regional and income-level lines became a feature of the Round as a whole. The Cairns Group was only one of several subgroups of contracting parties to play an active role in the negotiations. The growing involvement of countries other than the US and the EC in trade negotiations was evident even before the Round. One such group had its origins in the 1970s, when 'quadrilateral' talks were convened involving Japan and Canada as well as the EC and the US. The 'Quad' acted as an informal steering group during the Tokyo Round. During the Uruguay Round the Quad met twice every year, and was able to resolve a number of issues outside the main negotiations. The Quad countries agreed to consult more with other industrial countries, and met as a larger group on occasions. In addition, they have met as the 'Quint' (including Australia) when discussions were in the area of agriculture. Informal meetings of trade ministers had been held since 1983, involving about 25 countries, and supplementing the rare formal GATT ministerial meetings. This group played an important role in reviewing periodically the progress in the talks. An Asian–Pacific group had met occasionally since 1985 to discuss issues pertinent to that region. Later, after the breakdown of the talks in Montreal, the 'de la Paix Group', which included many of the countries that had taken part in the July 1986 informal session on the UR agenda, was formed and met regularly to maintain the momentum of the talks and develop proposals for many of the topics of the Round. Among other informal groups, the 'Rolle' group

submitted key ideas on trade in services; and the 'Morges Group' of ten countries particularly interested in agricultural trade met regularly during the Round to search for solutions to the outstanding issues in that area. This latter forum was mainly for senior officials to hold informal talks without the distraction of public scrutiny (Oxley, 1990, p. 159–60). The agricultural negotiations themselves were often among the 'Group of Eight', with the occasional involvement of a larger 'Group of 32' including more developing countries and others from the Morges Group. The Eight were Australia, Argentina, the EC, New Zealand, Finland, India, Japan and the US. Many other informal groups played a role in the negotiating process, as is usual on such occasions. The outcomes of such small group discussions were then relayed back to the formal negotiating groups. This wide participation in the process reflects both the increasing involvement of countries in trade discussions and a frustration of small and medium-sized countries that they have had too little voice in the formation of trade rules.

15. Financial Times, 16 December 1993, p. 1.
16. One such promise was to press for the inclusion of labour standards issues on the agenda of the WTO. In this the US was joined by France, but most other countries feared the opening of a new avenue for protectionism.
17. The Agreements went into effect for those members that had ratified them at that time. More countries followed over time. As of July 1995 there were over 100 signatories to the Agreements and hence members of the WTO.
18. The Declaration did, however, call for negotiations to 'take account of the approaches suggested in the work of the CTA'.
19. The US Special Trade Representative, Clayton Yeutter, and the Agriculture Secretary, Richard Lyng, are said to have agreed on the 'zero-option' on the principle that 'if you ask for a whole loaf you may get a half'. However, the subsequent defence of the request for the whole loaf became an objective in itself rather than a tactic to raise the eventual level of support reduction.
20. See Moyer and Josling (1990) for a discussion of the 1985 US Farm Bill.
21. Another technique surfaced in bilateral trade relations in the early 1990s, that of 'results-oriented' negotiations. Countries under such a scheme would negotiate on trade targets such as market shares, and then rely on selective intervention or liberalization to make those results come true. Market share negotiations had played a role in past agricultural trade talks, but were not advanced in the Uruguay Round.
22. An AMS did find its way into the US–Canada Free Trade Agreement as a device for opening up the cereals market. Non-tariff barriers between the US and Canada for grains could not be maintained by either party if the PSE was lower in the other country for that grain. In other words, non-tariff barriers could not be used in support of the more protectionist party's policies but could be kept by the less protectionist party. In the event, the non-tariff barriers were removed by the operation of this clause.
23. Historically, the EC had argued that agricultural protectionism was in principle acceptable as an exception to liberal rules of international trade. In the Punta del Este Declaration agriculture was to be integrated into the

trading system rather than isolated from it. The OECD ministers reinforced the message, and little was heard about special principles for agricultural trade after that time. Indeed, the EC itself abandoned much of the rhetoric on the need for special treatment for agriculture in its own 1987 proposal.

24. The cool reception among negotiators stemmed in part from a basic mistrust in US agricultural circles of the intentions of the EC Commission, and the fact that the signals in the paper were mixed. It also does not come easily for negotiators to praise each other's position papers. But it is tantalizing to think of what might have happened if the EC paper had been incorporated more into the mainstream of the talks rather than marginalized as 'more of the same old stuff'. With hindsight, one could argue that the EC suggestion that there could be bindings of 'maximum levels of support, protection and export compensation' foreshadowed the key elements of the final Agreement concluded six years later.

25. Not all was well in agricultural policy in the EC. The introduction of milk quotas in 1984, and their subsequent tightening in 1986, signalled a change in policy direction. The Commission had already proposed the goal of bringing EC cereal prices in line with those in the US. World prices were low, and budget costs high. But an advance obligation to remove price supports was difficult for European policy-makers to discuss, even if they privately may have wished for such an external discipline.

26. Though members of the Cairns Group, the Canadians were unconvinced that the Cairns Group could reach an agreement in time and produced their own paper to be on the safe side. This paper was somewhere between the US and Cairns Group in content. Canadian officials also floated the 'illustrative' proposal of a 50 per cent reduction in support levels over a ten-year period. This pace of reduction was not far out of line with the final outcome.

27. Subsequent to the defection of Denmark and the UK to the EC in 1973, bilateral trade pacts were negotiated to create a free-trade zone for manufactured goods which included both the EC and the EFTA. Agricultural trade had not been liberalized among the EFTA countries, and therefore was excluded from these bilateral packs. The levels of protection in the EFTA countries for agriculture were generally above those in the EC.

28. The countries concerned were Egypt, Jamaica, Mexico, Morocco and Peru. See GATT (1988a).

29. The negotiators apparently resorted to a thesaurus to find a word that would mean both reduction and elimination, without success.

30. The Agreement did include an 'intention to reduce support and protection levels for 1990 . . . either by using an AMS or by taking specific policy measures'. This presumably counts as the first incidence of the use of an AMS in a multilateral trade agreement (as opposed to a national proposal).

31. The change to tariffication had been foreshadowed in a US paper of November 1988, which for the first time in the Round mentioned it as a possible device (GATT, 1988e). This proposal was, however, still set in the context of a complete elimination of protection for agriculture. Governments would not maintain policies 'which would lead to a level of imports lower than the level which would occur in the absence of such

measures' (p. 2). The idea was to use the mid-term review to get agreement to a reform programme, with a new set of rules, including tariffication and the elimination of export subsidies, to be negotiated in January 1989.

32. This set of issues became known as 'rebalancing'. The clearest, and most controversial, example of the 'need' for rebalancing was the case of oilseeds. The EC had been operating a deficiency payment programme for oil-seeds, to allow for producer prices which could keep oil-seed production attractive to producers of cereals without increasing the low, bound tariffs on these products agreed to in the Dillon Round. Raising market prices for oil-seeds would help to increase demand for cereals, dairy products, and olive oil, all of which were in surplus. The Community sought to unbind the oil-seed tariffs as a way of solving the internal problem of cross-commodity price levels. Another important candidate for rebalancing was the whole group of 'cereal substitutes', such as manioc and corn gluten feed. The issue of rebalancing was linked to market access both by the fact that rebalancing would itself raise import barriers and also because the EC chose to insist on rebalancing as a precondition for its agreement on tariffication. The EC Commission asked a group of US and EC academics to study the issue of 'disharmonies' in agricultural policy, i.e. the problems caused by inappropriate relative policy price levels among closely related commodities (Koester et al., 1988), presumably with the hope of establishing that the problem was generic rather than an aberration caused by the politics of the CAP. Exporters remained steadfastly opposed to rebalancing, and it finally dropped from the negotiating table in the final stages of the Round.

33. GATT (1989c, p. 6). The 'corrective factor' proposal can be seen as foreshadowing the additional duty provisions of the Special Safeguard clause of the eventual Agreement.

34. It was never clear how one could convert deficiency payments to tariffs in the case of US cereals. The Commission appeared to argue at times that the US, by paying a deficiency payment, restricted potential imports of (European?) cereals. This was odd logic: it would have made more economic sense to have cast the US programmes as indirect export subsidies, but then the EC would have had to accept constraints on its own export subsidies in order to impose them on the US.

35. The AMS, however, was not to be based on a fixed reference price, as the Chairman's Draft (and the EC) had suggested.

36. GATT (1990d) (Hellström Draft). The Draft suggested a base period of 1990 (1988–90 average in the case of export quantities), as opposed to the EC suggestion of 1986. This would have considerably increased the depth of the cuts in support. In 1986 support was high because of low world prices. By 1990 world prices had recovered and support was consequently at a lower level.

37. CAP reforms until that time had been budget driven. The MacSharry reform was to cost even more money by putting more of the burden of income support on the taxpayer. The benefits from this type of reform are that it allows some more flexibility in trade policies by allowing domestic market prices to be reduced.

38. The status of these drafts was somewhat unclear. Were they a formal

offer or a tentative interpretation as to what might be implied by the proposed set of regulations? Countries clearly were not going to submit documents that contained elements with which they could not live. On the
other hand, the exercise was apparently meant more to determine the
implications of such an agreement rather than define the final position of
countries.

39. The case may also appear to be important because it seemed to demonstrate that there are effective limits to the extent to which countries can
grant domestic subsidies to their farmers, even though the GATT does
not directly constrain domestic subsidization. However, in this latter regard the ruling was not really quite as strong as one might have liked. By
concentrating attention on the *stability* of producer prices created by the
EC oil-seeds subsidy, the panel again managed to steer clear of the need
to rule on the acceptable *level* of the subsidy and the resulting producer
price. Had the EC paid its subsidy at a fixed rate, independent of the
current world price, the reasoning of the panel would not have worked.
Indeed, even if the EC subsidy had been double or triple its actual amount,
but invariable over time, the argument used by the panel would not have
sufficed to show that the EC subsidy had damaged US interests. Moreover, the panel ruling on the national treatment obligation was not fully
in line with economic logic. As the panel implied, if the subsidy paid to
crushers really were higher than the difference between EC target price
and world price, then crushers would have had an extra incentive to purchase EC oil-seeds. However, the panel failed to consider what the market implications would have been. With perfect competition among crushers,
their extra demand for EC oil-seeds would have driven up the domestic
price for oil-seeds to the point where crushers were indifferent between
buying EC oil-seeds and imported oil-seeds. Article III:4 would then not
have been violated, though nullification and impairment might have been
even more of an issue. On the other hand, the only market situation where
an excess subsidy to crushers could have existed without its resulting in
higher farm level prices in the EC is one in which there was imperfect
competition among crushers. Given the small number of companies involved, this was not an implausible state of affairs. In this case, crushers
would have deliberately held back their demand for EC-produced oil-seeds
in order to keep the price of their raw material low. Crushers would then
have collected an economic rent, but they would have purchased fewer
rather than more EC-produced oil-seeds. Under such circumstances it could
not really be said that domestic products were treated more favourably
than imported products.

40. The oil-seed dispute between the US and the EC had come to be en-
meshed in the Uruguay Round by an accident of timing and a linkage
with CAP reform. Putting the MacSharry payments in the Uruguay Round
green box would be tantamount to saying that the oil-seed payments were
also unobjectionable in the GATT. Clearly, the two issues had to be resolved together.

41. The press reports of the day speculated that the intervention by Delors
was on behalf of the French government. As Commissioners are meant to
be independent of national interests, this charge was denied. In the light

of the subsequent difficulties that French public opinion had with the deal, and his own interest in returning to French political life, it was reasonable to assume that Delors may have had something more in mind than the EC's desire to get an agreement with Washington to end the trade talks.

42. Countries were allowed to choose a different base year from which to start reductions in those cases where exports had increased significantly since the start of the Round. The final level of export subsidies allowed was not modified.

43. This is the wording which found its way into the Agreement on Agriculture.

44. By contrast, the last three years provided a case study in the political sensitivities of reaching an international agreement on agricultural issues. No new ideas surfaced over this period, and the quality of the debate deteriorated. Interest in the Round by economists waned, though analysts in national administrations were of course kept busy providing ammunition for the front line.

45. The situation in world agriculture was the subject of many academic and other papers over this period. See for instance Hathaway (1987). A brief statement of the academic consensus is found in the National Center for Food and Agricultural Policy (1988). This statement was authored by 'Twenty-Six Agricultural Trade Policy Experts from Eight Countries and Two International Organizations'.

46. The academic studies that demonstrated this externality include Tyers and Anderson (1988) and Roningen and Dixit (1990). The quantitative evidence is surveyed in Blandford (1990).

47. The search is well illustrated by the series of studies conducted under the auspices of the International Agricultural Trade Research Consortium, a group of academic and government economists that took a particular interest in the Round and the agricultural talks. See for instance Josling et al. (1989, 1990) and Magiera et al. (1990).

48. The level of trade distortion caused by, and the type of instrument used in, policy are not unconnected. It is a feature of domestic farm policies that if price levels are kept high relative to those on world markets, all instruments will appear harmful to international trade, whereas if prices are kept low then almost any type of policy is relatively trade-neutral. The choice between rules and reductions as a way of blunting an objectionable policy may have much to do with the ease of implementing an agreement. Rules need credible dispute settlement procedures, threats of sanctions, and general agreement among most countries that the rule is sensible. Reductions need quantification of the policy instrument or its effect, monitoring of the level over time and a belief that most countries will not try to evade the intention of the reduction. Moreover, rules can be translated quickly into domestic legislation along with the rest of the results from a trade negotiation whereas reduction commitments may need periodic domestic action over an extended period of time.

49. Implications for domestic policy presumably surface later, when governments have to put into operation what they have agreed in the GATT. Often, in agriculture, the domestic implications weaken the resolve of governments to implement decisions taken in the GATT, leading to a lack of credibility.

50. It was never clear that detailed policy plans, as suggested by both the US and the Cairns Group in their original proposals, could ever have been agreed for each country in the GATT.
51. It had also not escaped the notice of the Community that the OECD PSE for Canada and the US showed them to have more agricultural protection than appeared from the rhetoric of their negotiating positions.
52. Such distinctions were already being discussed between red-, amber- and green-light policies in the 'traffic light system' of the subsidy negotiations. In agriculture, the traffic light metaphor gave way to red, amber, blue, and green 'boxes' into which policy instruments would fall.
53. See, for instance, the discussion in Tangermann, Josling, and Pearson (1987).
54. The new rules, as well as the nature of reduction commitments, will be described in more detail in the following chapter.
55. Ironically, it was Chile, a member of the Cairns Group and a leading advocate of liberal trade policies, which introduced a 'price band' to protect its own agricultural import sectors against world price fluctuations. This type of policy has since spread to other countries in Latin America.
56. The Canadian concern centred on the issue of whether control of domestic milk production allowed import quotas on ice cream and yoghurt. A GATT panel ruled that these were not 'like products', and thus threatened to undermine control of the dairy product market. The Canadian 'clarification' would have removed this constraint on the dairy policy and made it easier to impose quantitative restrictions on imports.
57. For a discussion of the merits of tariffication which emerged during the Round, see Zeitz and Valdés (1988). The issue of tariffication as endorsed by the US and the Cairns Group proposals of 1989 is discussed in Josling et al. (1989).
58. The rules for tariffication, as well as the numerical targets (for developed countries a 36 per cent decline over a six-year period, with a minimum of 15 per cent per tariff line; for developing countries a 24 per cent decline over ten years) are contained in GATT (1993).
59. See Chapter 8 for more details on the resulting tariff schedules, and Chapter 9 for an assessment of the task of reducing these barriers in future negotiations.
60. The 'red light' category of banned domestic subsidies was dropped from active discussion in early 1990.

8 THE URUGUAY ROUND AGREEMENT ON AGRICULTURE

1. The most convenient publication of the legal text which resulted from the UR, including the text of the GATT 1947, is WTO (1995a). In the following pages, all citations from the UR agreements are from this publication.
2. In the following, if the term 'Agreement' is used without further specification it refers to the Agreement on Agriculture.
3. The full title of the working document is *Modalities for the Establishment of Specific Binding Commitments under the Reform Program: Note by the Chairman of the Market Access Group* (GATT, 1993). In the following, this document will be referred to as Modalities. Nearly all of the

text of the Modalities is taken directly from the Dunkel Draft (GATT, 1991, *Draft Final Act* . . .), where it formed Part B of the proposed agreement on agriculture. After the end of the UR negotiations, the Modalities document completely lost any legal standing, and its text explicitly says that it shall not be used for dispute settlement proceedings under the WTO.

4. A summary evaluation of the Agreement on Agriculture, including the implications for a number of major countries, is provided in Josling *et al.* (1994). Provisions of the Agreement are assessed in detail in Tangermann (1994). The following analysis draws heavily on these two publications.

5. In the following, the term 'Article', if not specified otherwise, always refers to an article of the Agreement on Agriculture. With the establishment of the WTO, the term 'Member' (of the WTO) has replaced the term 'contracting party' used in the GATT.

6. Because all Articles of the GATT 1947 have become part of the GATT 1994, unchanged, they will in the following be referred to as articles of the 'GATT', without a year added.

7. According to the footnote, NTM can, however, be maintained if they are allowed under other general, non-agriculture-specific provisions of the GATT, for example balance of payments provisions, or any of the new agreements on trade in goods.

8. Criteria governing the calculation of the tariff equivalent (also to be based on the 1986–8 reference period) for these cases are specified in an attachment to Annex 5. They closely resemble the respective provisions in the Modalities. It is interesting to note that this is the only place where criteria for how to establish commitments have entered the Agreement, while everywhere else they have remained in the Modalities. It is said that an attempt was made to convince Japanese negotiators to establish the tariff equivalent for rice before the end of the UR negotiations, such that this attachment would have been unnecessary. However, Japan refused to do so, and even declined the suggestion to submit the tariff equivalent in a sealed envelope to the Director-General of the GATT. Thus, there was no way around including conditions for the calculation of tariff equivalents in the Agreement, just for the rice clause.

9. All reduction rates reported here and in the following are those relating to developed countries. For developing countries, reduction rates are two thirds of those for developed countries, and the implementation period is ten years rather than six years. The least developed countries are exempt from reduction commitments (Modalities, paragraphs 15 and 16).

10. Where actual imports during the base period were less than existing quotas, base period quotas were to be maintained.

11. The origins of tariff-rate quotas can, though, be traced by comparing Schedules with the respective supporting tables included in the offers for the verification process. Also, if a tariff-rate quota in a Schedule is designated to be allocated to given exporting countries, its origin is a current access commitment (but the reverse is not necessarily true).

12. Economic theory shows that in certain cases it is impossible to establish the tariff which is equivalent to a given non-tariff measure: see Corden (1984, pp. 79–81). However, this problem was not, and probably could not be, reflected in the Modalities.

13. However, the products that have undergone tariffication can generally still be identified because most of them are designated, in a separate column of the Schedule, as being subject to the Special Safeguard Provision (which is available only where tariffication took place).

14. In some cases where important trade concerns were involved, negotiations took place over the tariff equivalents to be established. Usually these negotiations were pursued before the verification process.

15. For the implications of this commitment, and some significant problems which can arise from it, see Tangermann and Josling (1994) and Josling and Tangermann (1995). The EC made a similar commitment for rice.

16. Another instructive example of the importance of the base period data is the calculation of the tariff equivalent of the US Meat Import Law. Australia and New Zealand as beef exporters wanted the US to use a higher external reference price than initially suggested, so that the resulting tariff equivalent would be low. The irony was not lost on the participants. Earlier, in the OECD, Australia and New Zealand had urged the US to accept a low external reference price in the PSE calculations to signify how high US protection was for beef under the same policies. In the end the somewhat high tariff equivalent was agreed. See Josling *et al.* (1994, p. 29).

17. In the tariffication process, countries were free to bind either specific or *ad valorem* duties. All sorts of specifications can be found in the Schedules. For example, in some cases, combinations of specific and *ad valorem* duties have been bound. In other cases *ad valorem* duties have been bound, with a given minimum or maximum specific duty, etc. Also, countries could choose the currency in which to bind a specific duty (see Article II:6 of the GATT). For example, Poland has bound duties in ECU, generally at the level of EC tariffs. To make tariffs comparable for the table shown here, all tariffs have been expressed in their *ad valorem* equivalents, as a percentage of world market prices in 1992–3.

18. Cases in point are EC tariffs for oats and grain-based livestock: see Tangermann (1995) for details. Countries have not always made full use of the bound rates, and in some cases tariffs actually applied at the beginning of the implementation period were significantly below the bound rates.

19. For an analysis of the implications of unbalanced rates of protection in the EC and the US, see Koester *et al.* (1988).

20. World market prices used for converting specific tariffs into *ad valorem* equivalents are 1993 unit values of EC imports. For more detail, see Tangermann (1995).

21. In many cases, several different tariff items have one joint quota, with tariffs differing among items, such that the actual number of products that can be imported under tariff-rate quotas is much higher than the number of quotas.

22. These preferences are part of the Association Agreements with countries from Central Europe ('Europe Agreements'). As these agreements were concluded after the base period for the UR Agreement on Agriculture, they do not come under the current access provisions.

23. This is the case in the EC, for example, for meat, wheat, and maize.

Another example is the tariff-rate quota for poultry meat which the EC opened up as a result of the bilateral negotiations with the US in the context of settling the oil-seeds dispute. In the EC Schedule, this quota is also listed under minimum access.

24. For example, it is believed that in some cases importing countries have promised to import agreed shares of their minimum access quantities from given exporters, in spite of the requirement to implement the tariff-rate quotas on a MFN basis.

25. For example, if it were the case that imports from a given country always arrive during the first half of the week, it would not be acceptable to set tariff rates such that they are higher from Monday to Wednesday than during the rest of the week.

26. When the implementation period began, the EC adopted cereal duties which differentiated between quality categories, but not between individual shipments. However, the USA and Canada were not entirely satisfied with that regime and reserved the right to challenge it in the WTO.

27. According to the definition of terms in Article 1 of the Agreement, 'outlay' includes revenue forgone.

28. For developing countries, subsidies on costs of export marketing, as well as those on internal transport costs of export shipments, do not fall under the reduction commitment during the implementation period (Article 9(4)).

29. Through the back door of this part of the flexibility provision, the text of the Agreement implicitly contains the rates of reduction agreed for export subsidies. This contrasts with the cases of market access and domestic support where reduction rates are found only in the Modalities.

30. The EC itself used the front-loading provision for other products as well, to raise the quantities of allowable subsidized exports by 13.2 per cent for poultry, 5 per cent for cheese, 0.7 per cent for other dairy products, 3 per cent for eggs and 21.2 per cent for tobacco. The USA has also made use of this provision, raising allowable subsidized exports, for example, by 262.2 per cent for rice, 141.9 per cent for eggs, and 132.2 per cent for vegetable oils.

31. Typically, the export subsidy section of a country Schedule has 22 lines for the 22 individual product groups, plus one additional line for incorporated products, with that line having only expenditure commitments for each year of the implementation period.

32. In 1985, EC Commissioner for External Affairs Frans Andriessen promised, in a letter to the Australian government, that the EC would not subsidize beef exports to countries in East Asia to which the EC did not at the time export and to which Australia had in recent years had substantial beef exports. The EC honoured this commitment, and reaffirmed it in 1991.

33. Guarding against this type of policy substitution was one of the major aims of the original suggestion to bind commodity-specific AMS levels. Of this suggestion, only the provision in the peace clause, according to which subsidies are no longer exempt from the GATT challenge if their commodity-specific level exceeds that decided in the 1992 marketing year, has remained.

34. To add another twist to this example, assume that the government of the country concerned lets farmers believe that above-quota production may

be honoured when it comes to a reallocation of the production quota among farmers. Is the indirectly promised quota rent in the future then a form of export subsidy?

35. For developing countries, the *de minimis* percentage was set at 10 per cent.
36. The following discussion focuses only on selected conceptual differences. There are many other differences between PSE and AMS which will not be mentioned here because they are not central to the argument.
37. For the AMS, green box subsidies would not, of course, be included.
38. There is, though, one particular factor which distinguishes the (implicit) constraint on border measures resulting from the AMS from the more specific commitments on border measures. The AMS uses a fixed external reference price in domestic currency, while both tariffs and export subsidies relate to actual world market prices. This issue will be taken up in Section 6.
39. The distinction between trade-distortion effects and production effects is interesting to note. One wonders whether there are any measures that have production effects but do not distort trade. Moreover, if there are any such measures, why should the GATT want to regulate them?
40. However, Article 18:4 of the Agreement provides that in reviewing the implementation of the Agreement (in the Committee on Agriculture) 'due consideration [shall be given] to the influence of excessive rates of inflation on the ability of any Member to abide by its domestic support commitments'. It will be interesting to see how the Committee will deal with this provision. In this context it is also interesting to note that countries could express the AMS commitments in a foreign currency. Poland, for example, has expressed its AMS commitment in US dollars, thereby avoiding pressure that might have resulted from zloty inflation.
41. Issues related to the treatment of sanitary and phytosanitary measures in the GATT are discussed in more detail by Kramer (1988), Bredahl and Forsythe (1989), Petrey and Johnson (1993), and Wirth (1994).
42. The USA, where growth hormones are permitted, considered the EC ban not to be supported by scientific evidence and claimed, contrary to the EC, that there was no risk to human health from consumption of hormone-treated beef. The EC disagreed. For an analysis of the controversy see Halpern (1989).
43. In an interesting, though somewhat ambiguous, clause the Agreement on SPS suggests that in developing guidelines for the practical implementation of risk assessment 'the exceptional character of human health risks to which people voluntarily expose themselves' shall be taken into account.
44. Not only 'identical' (as in Article XX) but also 'similar' conditions must be treated alike, and the injunction of no discrimination 'between countries' now explicitly includes the territory of the country applying the SPS.
45. Section B of the GATT Article XVI, consisting of Article XVI:2–5, was added in 1955. Before this amendment there was no explicit prohibition of export subsidies in the GATT. The US proposals for the Havana Charter had contained provisions similar to those regarding export subsidies added to Article XVI in 1955, including the tolerance of agricultural export subsidies as long as they did not result in a more than equitable

share of world trade, but these were not carried into the GATT in 1947. See Chapters 1 and 2.

46. Note, however, that reduction commitments relate to domestic *support*, rather than to domestic *subsidies*. As discussed above, this means that in many cases domestic subsidies in the narrow sense may not really have to be reduced.

47. The colour spectrum of subsidies is even richer if direct payments under 'production-limiting programmes' are called 'blue' as has become customary.

9 THE FUTURE FOR AGRICULTURE IN THE GATT

1. Some of this protection crept in during the process of tariffication: most of it was hidden in the undergrowth of non-tariff measures.

2. Since the 1992 Maastricht Treaty it has become customary to speak of the European Union, rather than the European Community. As the present chapter is mainly forward looking, the acronym EU is used instead of EC.

3. Canada, for instance, has been redesigning its safety-net programmes for agricultural incomes to avoid the incentive to produce. Changes in Canadian supply management and grain transportation subsidies are also examples of the influence of the ideas from the Round on domestic policy formation.

4. At the time of writing (September 1995) world cereal prices are again high and stocks low. Any policy-induced expansion is likely again to hit a weakening market in two or three years' time. It will be an interesting test of the new approach to agricultural policy to see whether the high prices are seen as an opportunity to ratchet up support at low current budget cost or a chance to remove such support with little impact on current farm incomes. Is it farmers or governments (or taxpayers) that will absorb the downside price risk?

5. Whether it is easier to remove import protection when prices are high, and therefore protection is less needed, or when prices are low and hence protection is more noticeable, is not clear. It is probably best to get political agreement on the agenda when the problems are most apparent and negotiate the detailed commitments when the price impacts are likely to be less noticeable. This in effect happened in the Uruguay Round. Weak prices until 1987 strengthened countries' resolve and firm prices from 1988–92 helped to ease the process of establishing the tariff cuts.

6. Restrictive rules, banning or prohibiting actions, do not need to be commodity specific and do not require quantification. Permissive regulations, which allow an action but then attempt to put limits on it, have proved to require much more specific quantification.

7. Why tariffication should have been possible in the Uruguay Round and not before is itself an interesting question for research. The EC signified early on in the Round that it was prepared for some binding of support protection and export compensation. The eventual deal allowed the EC to keep stabilizing elements of the variable levy. Thus one could say that the turning point came at that stage in the Round when the EC decided that it could preserve its stabilization policies and exchange its use of the

variable levy for less onerous restrictions on export subsidies.
8. The Committee is set up in Article 18 of the Agreement. The issue of future negotiations is dealt with in Article 20, entitled Continuation of the Reform Process.
9. The 'water' in a tariff is the unused protection when imports are in effect prohibited. The 'dirty' element in the agricultural tariffs refers to the use of price gaps between domestic and world markets which overstated the existing protection. Tariff bindings can also be above the actual tariff in operation, giving an element of discretion to governments. Thus a reduction in the high rates of tariff removes the water, cleans up the tariff and removes the discretionary element of ceiling bindings.
10. This was the case, for example, in the EU, Japan, and the USA. See Chapter 8.
11. There would of course be a danger in announcing such a scheme in advance. Countries might choose to raise tariffs to their bound levels to avoid the cut.
12. It would not of course remove the element of protection introduced through 'dirty tariffication', and there could still be 'water' in the tariffs as currently applied. It would only remove the discretionary element of protection that countries were able to build in for their own flexibility. This would, however, increase considerably the credibility of the liberalization process. It would also put a limit to the spread of 'price band' systems of variable protection that several Latin American countries have adopted to stabilize domestic prices.
13. The experience with the textile quotas in the Multifibre Arrangement provides an object lesson. These quotas started out as a way of liberalizing trade and ended up by being restrictive. If a quota is not liberalized at a rate faster than the market would otherwise have expanded then it becomes more restrictive over time. However, the tariff rate quotas agreed in the UR are less dangerous than quotas under the 'old' Multifibre Agreement as above-quota imports are possible under the newly bound tariffs in agriculture.
14. For a fuller discussion of these issues see Josling (1993).
15. The GATT outcome should make negotiation of future regional trade agreements much easier. For example, large parts of the GATT text on subsidies were included in the NAFTA. These would have had to be negotiated separately if there were no GATT deal.
16. MERCOSUR, the newly formed Common Market of the Southern Cone, includes three major agricultural exporters, Argentina, Brazil, and Uruguay as well as Paraguay. Agricultural trade is included in the free-trade provisions, and increased agricultural exports, particularly from Argentina to Brazil, have followed the reduction of trade barriers.
17. France and the Netherlands were well aware of the importance of regional markets and insisted that agricultural trade be freed within the European Economic Community from its inception. One should, however, add the caveat that currency instability prevented for over twenty-five years the creation of a truly free internal market in agricultural products. Border taxes (monetary compensatory amounts) were levied and subsidies paid on agricultural trade from 1969 to 1994, to offset the effect of

currency movements. Moreover, the high level of external protection implied that much of the extra internal trade was at the expense of cheaper sources from third countries.

18. All domestic production that would otherwise have gone for domestic consumption could be exported to the partner even with the strictest rules of origin. If domestic consumption in one country is large relative to the imports into the partner country, its tariff level, if lower, will tend to dominate.

19. If all these plans materialize, the US will be a 'mega-hub' with spokes reaching to the EU, the APEC countries and the Americas. The EU and Asia will no doubt find it in their interest to avoid the negative implications of this by negotiating an agreement on trade between themselves. Moreover, by that time the multilateral system will have had time to come to terms with these supra-regional blocs – perhaps by incorporating their timetables into a global plan for trade liberalization.

20. As an analytical problem, the more interesting issue is whether 'free trade' implies a disregard for environmental policies. The answer is not difficult to state. Countries should ensure that they are correctly managing their own environmental affairs so that the polluting industry 'pays' for costs imposed on society, whether from producing goods for the domestic market or for export. Free trade then becomes entirely consistent with domestic attempts to control pollution. The same argument holds regarding the potentially negative environmental effects of international transport resulting from expanded trade. If governments make sure that resources used in transport activities are properly priced then externalities involved in transport will be taken into account when trading activities take place.

21. This is indeed the view which is reflected in the Rio Declaration on Environment and Development (UN, 1992).

22. CITES directly bans trade in endangered species, and the Montreal Protocol restricts trade with non-signatories. See Esty (1994, p. 130).

23. Another type of problem is a direct bilateral externality (one country emits waste into a river flowing into another country's territory). Trade policies may be justified in that situation if other ways of solving the problem cannot be found. However, it would probably be difficult to establish GATT-legality of such measures.

24. This was recognized in the Rio Declaration of 1992.

25. The admission of China into the WTO will immediately raise the problem of state trading in agricultural products in a way that can no longer be ignored. Discussions on the terms of entry (or re-entry) of China into the mainstream of the trade system have already stumbled on this issue.

26. This issue has dogged US–Canadian discussions over the activities of the Canadian Wheat Board.

27. This problem is of course not new. The Havana Charter had a section on restrictive trade practices which anticipated some of the problems that would be faced in a world of international investment, powerful traders, and expanding intra-firm trade across borders. The UNCTAD has over the years also attempted to deal with some of these issues.

References

Anderson, K. (1992), 'The standard welfare economics of policies affecting trade and the environment', in: K. Anderson and R. Blackhurst (eds), *The Greening of World Trade Issues* (New York: Harvester Wheatsheaf).

Atlantic Council (1973), *U.S. Agriculture in a World Context* (Washington, DC: Atlantic Council).

Australian Industries Assistance Commission (1980), *Annual Report 1979–80* (Canberra: AIAC).

Bergsten, C.F. (1971), *Reshaping the International Economic Order* (Washington, DC: Institute for International Economics).

Bergsten, C.F. (ed.) (1973), *The Future of the International Economic Order: An Agenda for Research* (Lexington: Lexington Books).

Bergsten, C. F. (1975), *Towards a New International Economic Order: Selected Papers of C. Fred Bergsten, 1972–1974* (Lexington: Lexington Books).

Blandford, D. (1990), 'The costs of agricultural protection and the difference free trade would make', in: F. Sanderson (ed.), *Agricultural Protectionism in the Industrialized World* (Washington, DC: Resources for the Future).

Bredahl, M.E., and K.W. Forsythe (1989), 'Harmonizing phyto-sanitary and sanitary regulations', *World Economy*, vol. 12, pp. 189–206.

Brookings Institution (1973), *Toward the Integration of World Agriculture: A Report by 14 Experts from North America, the European Community and Japan* (Washington, DC: Brookings Institution).

Brown, W.A., Jr. (1950), *The United States and the Restoration of World Trade* (Washington, DC: Brookings Institution).

Brunthaver, C.G. (1965), 'UK Grain Agreement: Format for an International Grain Agreement?' *Journal of Farm Economics*, vol. 47, no. I, pp. 51–9.

Butz, Earl L. (1974), *Statement of the US Secretary of Agriculture before the Senate Finance Committee* (Washington, DC: US Department of Agriculture).

Camps, M. (1964), *Britain and the European Community, 1955–1963* (London: Oxford University Press).

Camps, M. (1965), *What Kind of Europe? The Community since de Gaulle's Veto* (London: Oxford University Press).

Camps, M. (1966), *European Unification in the Sixties* (New York: McGraw-Hill for the Council on Foreign Relations).

Camps, M. (1981), *Collective Management: The Reform of Global Economic Organizations* (New York: McGraw-Hill for the Council on Foreign Relations).

Casadio, G. P. (1973), *Transatlantic Trade: USA–EEC Confrontation in the GATT Negotiations* (Lexington: Westmead).

Cline, W.R. (ed.) (1983), *Trade Policy in the 1980s* (Washington, DC: Institute for International Economics).

Committee on Economic Development (1971), *The United States and the European Community: Policies for a Changing World* (Washington, DC: US Government Printing Office).

275

Coombes, D. (1970), *Politics and Bureaucracy in the European Community* (London: George Allen and Unwin for Political and Economic Planning).

Corbet, H. (1985), 'Agricultural priorities after the Tokyo Round negotiations', in: H. de Haen, G. L. Johnson and S. Tangermann (eds), *Agriculture and International Relations, Analysis and Policy: Essays in Memory of Theodor Heidhues* (London: Macmillan; New York: St. Martin's Press).

Corbet, H. and R. Jackson (eds) (1974), *In Search of a New World Economic Order* (London: Croom Helm for Trade Policy Research Centre).

Corden, W.M. (1984), 'The normative theory of international trade', in: R.W. Jones and P.B. Kenen (eds), *Handbook of International Economics*, vol. I (Amsterdam: North-Holland).

Council of the European Communities (1973), *The Development of an Overall Approach to the Forthcoming Multilateral Trade Negotiations in GATT*, I/135/73 COMER 42, Luxembourg, June 29.

Croome, J. (1995), *Reshaping the World Trading System: A History of the Uruguay Round* (Geneva: World Trade Organization).

Cuff, R.L. and J.L. Granatstein (1977), 'The rise and fall of Canadian free trade 1947–8', *Canadian Historical Review*, vol. 58, no. 4, pp. 459–82.

Curtis, Th.B. and J.R. Vastine, Jr. (1971), *The Kennedy Round and the Future of American Trade* (New York: Praeger).

Curzon, G. (1965), *Multilateral Commercial Diplomacy: The General Agreement on Tariffs and Trade and Its Impact on National Commercial Policies and Techniques* (London: Joseph).

Curzon, G. and V. Curzon (1976), 'The management of trade relations in the GATT', in: A. Shonfield (ed.), *International Economic Relations of the Western World 1959–1971*, vol. 1. (London: Oxford University Press for the Royal Institute of International Affairs).

Dahrendorf, R. (1974), 'External relations of the European Community', in: H. Corbet and R. Jackson (eds), *In Search of a New World Economic Order* (London: Croom Helm for Trade Policy Research Centre).

Dam, K.W. (1970), *The GATT: Law and International Economic Organization* (Chicago: University of Chicago Press).

Davey, W.J. (1993), 'The rules for agricultural trade in GATT', in: M. Honma, A. Shimizu and H. Funatsu (eds), *GATT and Trade Liberalization in Agriculture* (Otaru: Otaru University of Commerce).

de C. Grey, R. (1983), 'A note on US trade practices', in: W.R. Cline (ed.), *Trade Policy in the 1980s* (Washington, DC: Institute for International Economics).

Destler, I.M. (1986), *American Trade Politics: System Under Stress* (Washington, DC: Institute for International Economics; New York: The Twentieth Century Fund).

Deutsche Bundesbank (1994), *Statistisches Beiheft zum Monatsbericht, Reihe 5: Devisenkursstatistik*.

Diebold, W. (1952), *The End of the ITO* (Princeton: Princeton University Press).

Djonovich, D.J. (ed.) (1978), *United Nations Resolutions*, Series 1, Vol. XIV, 1972–1974 (New York: Oceana Publications) Res. No. 3201, pp. 527–9.

EC Commission (1992), *Agriculture in the GATT Negotiations and the Reform of the CAP*, Communication from the Commission, Document SEC(92) 2267 final, Brussels, 25 November.

References 277

Eminent Persons Group on World Trade (1990), *Meeting the World Trade Deadline: Path to a Successful Uruguay Round* (No location, no publisher, July).

Esty, D.C. (1994), *Greening the GATT: Trade, Environment and the Future* (Washington, DC: Insitute for International Economics).

EUROSTAT (1994), *Intra- und Extra-Handel der EU*, CD ROM version (Luxemburg).

Evans, J.W. (1971), *The Kennedy Round in American Trade Policy. The Twilight of the GATT?* (Cambridge: Harvard University Press).

FAO (1954), *Disposal of Agricultural Surpluses: Principles Recommended by FAO* (Rome: FAO).

FAO (1973), *Agricultural Protection: Domestic Policy and International Trade*, C 73/LIM/9.

FAO (1975), *Agricultural Protection and Stabilisation Policies: A Framework for Measurement in the Context of Agricultural Adjustment*, C 75/LIM/2.

FAO (1995), *Commodity Review and Outlook, 1994–95* (Rome: FAO).

Flanigan, P.M. (1973), *Agricultural Trade and the Proposed Round of Trade Negotiations*, 93rd Congress, 1st Session, 30 April (Washington, DC: US Government Printing Office for the US Senate Committee on Agriculture and Forestry).

Fraser, G.O. (1975), 'U.S. agriculture's stake in the world trade negotiations', *Foreign Agriculture* (US Department of Agriculture), vol. 13, no. 7, pp. 2–4.

Gardner, R.N. (1980), *Sterling–Dollar Diplomacy in Current Perspective: The Origins and Prospects of Our International Economic Order* (New York: Columbia University Press).

GATT (1950), *The Use of Quantitative Restrictions for Protective and Commercial Purposes*, GATT/1950–3, July (Geneva).

GATT (1955), *Commodity Problems: Interim Reports*, L/320, 11 February, Annex A.

GATT (1958), *Trends in International Trade: A Report by a Panel of Experts* (Haberler Report), October (Geneva).

GATT (1962a), *Trends in Agricultural Trade: Report of Committee II on Consultations with the European Economic Community*, December.

GATT (1962b), *Trade in Agricultural Products: Reports of Committee II on Country Consultations*, mimeo, June.

GATT (1963), Press Release 751, 17 May.

GATT (1964a), *Statement by the Representative of the European Community before the GATT Committee on Agriculture Regarding the Negotiating Plan of the EEC for the Agricultural Part of the Kennedy Round*, TN 64/AGR/1, 19 February.

GATT (1964b), *Contribution of the European Community Relating to the Negotiation on Agricultural Products in the GATT Trade Negotiations*, TN 64/AGR/5, 3 August.

GATT (1965), *Proposals on the Negotiations on Agriculture*, TN/64/AGR/W.1, 27 January.

GATT (1967), *Legal Instruments Embodying the Results of the 1964–1967 Trade Conference*, Vol. V, pp. 3677–91.

GATT (1978), *Statement by Several Delegations on Current Status of the Tokyo Round Negotiations*, MTN/INF/33, 14 July.

GATT (1979), *The Tokyo Round of Multilateral Trade Negotiations* (Report by the Director-General of the GATT), GATT/1234, 12 April.

GATT (1980), *The Tokyo Round of Multilateral Trade Negotiations, II: Supplementary Report.*

GATT (1982a), *GATT Activities in 1981.*

GATT (1982b), *Agriculture in the GATT*, CG 18/W/59.

GATT (1983), *European Economic Community: Subsidies on Wheat Flour: Report of the Panel*, SCM/42, 21 March.

GATT (1984), *Draft Elaboration*, AG/W/9, 25 June.

GATT (1986), *Draft Elaboration*, AG/W/9 Rev. 3, 4 June.

GATT (1987a), *US Proposal to the Uruguay Round Negotiating Group on Agriculture*, MTN.GNG/NG5/W/14, 7 July.

GATT (1987b), *EC Proposal to the Uruguay Round Negotiating Group on Agriculture*, MTN.GNG/NG5/W/20, 26 October.

GATT (1987c), *Cairns Group Proposal to the Uruguay Round Negotiating Group on Agriculture*, MTN.GNG/NG5/W/21, 26 October.

GATT (1987d), *Japanese Proposal for Negotiations on Agriculture*, MTN.GNG/NG5/W/39, 26 December.

GATT (1987e), *Proposal by the Nordic Countries (Finland, Iceland, Norway and Sweden)*, MTN.GNG/NG5/W/35, 1 December.

GATT (1988a), *News of the Uruguay Round of Multilateral Trade Negotiations*, no. 19, October.

GATT (1988b), *News of the Uruguay Round of Multilateral Trade Negotiations*, no. 16, 31 May.

GATT (1988c), *News of the Uruguay Round of Multilateral Trade Negotiations*, no. 18, 2 August.

GATT (1988d), *News of the Uruguay Round of Multilateral Trade Negotiations*, no. 20, 4 November.

GATT (1988e), *A Framework of Agricultural Reform Submitted by the United States*, MTN.GNG/NG5/W/83, 11 November.

GATT (1989a), *Trade Negotiations Committee: Mid Term Meeting*, MTN.TNC/11, 21 April.

GATT (1989b), *Submission of the United States on Comprehensive Long-Term Agricultural Reform*, MTN.GNG/NG5/W/118, 25 October.

GATT (1989c), *Global Proposal of the European Community on the Long-Term Objective for the Multilateral Negotiation on Agricultural Questions*, MTN.GNG/NG5/W/145, 20 December.

GATT (1989d), *Comprehensive Proposal for the Long-term Reform of Agricultural Trade: Submission by the Cairns Group*, MTN.GNG/NG5/W/128, 27 November.

GATT (1989e), *Submission by Japan*, MTN.GNG/NG5/W/131, 6 December.

GATT (1990a), *Framework Agreement on Agricultural Reform Programme: Draft Text by the Chairman*, MTN.GNG/NG5/W/170, 11 July.

GATT (1990b), *A Proposal for the Agricultural Negotiations Submitted by the United States*, October.

GATT (1990c), *European Community Offer in Agriculture*, November.

GATT (1990d), *Elements for Negotiation of a Draft Agreement on the Agricultural Reform Programme* (Hellström Draft), December.

GATT (1991), *Draft Final Act Embodying the Results of the Uruguay Round of Multilateral Trade Negotiations* (Dunkel Draft), MTN.TNC/W/FA, 20 December.

GATT (1993), *Modalities for the Establishment of Specific Binding Commitments under the Reform Programme: Note by the Chairman of the Market Access Group*, MTN.GNG/MA/W/24, 20 December.

GATT (1994), *Schedules of Market Accession Concessions*, Electronic version.

GATT Secretariat (1993), *An Analysis of the Proposed Uruguay Round Agreement, with Particular Emphasis on Aspects of Interest to Developing Countries*, MTN.TNC/W/122, MTN.GNG/W/30, 29 November.

GATT Secretariat (1994), *The Results of the Uruguay Round of Multilateral Trade Negotiations, Market Access for Goods and Services: Overview of the Results*, November.

GATT, *Basic Instruments and Selected Documents*. (BISD). Various issues.

Golt, S. (1974), *The GATT Negotiations, 1973–75: A Guide to the Issues* (London: British–North America Committee).

Halpern, A.R. (1989), 'The US–EC hormone beef controversy and the Standards Code: Implications for the application of health regulations to agricultural trade', *North Carolina Journal of International Law and Commercial Regulation*, vol. 14, pp. 135–55.

Hardin, C.M. (1971), 'Needed: A market-oriented agricultural trading world', in: Williams Commission, *United States International Economic Policy in an Interdependent World*, Report to the President submitted by the Presidential Commission on International Trade and Investment Policy (Washington, DC: US Government Printing Office).

Harris, S. (1977), *EEC Trade Relations with the USA in Agricultural Products*, Occasional Paper 3 (Ashford: Centre for European Agricultural Studies, Wye College).

Hartwig, B. (1992), 'Die GATT-Regeln für die Landwirtschaft: Eine Ökonomische Analyse ihrer Wirksamkeit vor dem Hintergrund der Streitbeilegung', *Agrarwirtschaft*, Sonderheft 134.

Hartwig, B. and S. Tangermann (1987), *Legal Aspects of Restricting Manioc Trade Between Thailand and the EEC* (Kiel: Wissenschaftsverlag Vauk).

Hathaway, D.E. (1983), 'Agricultural trade policy for the 1980s', in: W.R. Cline (ed.), *Trade Policy in the 1980s* (Washington, DC: Institute for International Economics).

Hathaway, D.E. (1987), *Agriculture and the GATT: Issues in a New Trade Round* (Washington, DC: Institute for International Economics).

Hauser, H. and K.-U. Schanz (1995), *Das neue GATT: Die Welthandelsordnung nach Abschluß der Uruguay-Runde* (Munich and Vienna: Oldenbourg).

Hedges, I.R. (1967), 'Kennedy Round agricultural negotiations and the World Grains Arrangement', *Journal of Farm Economics*, vol. 49, no. 5, pp. 1332–41.

Hudec, R.E. (1975), *The GATT Legal System and World Trade Diplomacy* (New York: Praeger).

Hudec, R.E. (1987), 'Transcending the ostensible: Some reflections on the nature of litigation between governments', *Minnesota Law Review*, vol. 72, no. 10, pp. 111–13.

Hudec, R.E. (1993), *Enforcing International Trade Law: The Evolution of the Modern GATT Legal System* (Salem, NH: Butterworth).

Hufbauer, G.C. (1983), 'Subsidy issues after the Tokyo Round', in: W.R. Cline (ed.), *Trade Policy in the 1980s* (Washington, DC: Institute for International Economics).

Hufbauer, G.C. and J.J. Schott (1985), *Trading for Growth: The Next Round of Trade Negotiations* (Washington, DC: Institute for International Economics).

IMF (1994), *International Financial Statistics Yearbook* (Washington, DC: IMF).

Ingersent, K.A., A.J. Rayner and R.C. Hine (eds) (1994), *Agriculture in the Uruguay Round* (London: Macmillan).

Jackson, J.H. (1969), *World Trade and the Law of GATT* (Charlottesville: Michie).

Jackson, J.H. (1978), 'The crumbling institutions of the liberal trade system', *Journal of World Trade Law*, vol. 12, pp. 93–106.

Johnson, D.G. (1970), in: *A Foreign Economic Policy for the 1970s*, Hearings before the Subcommittee on Foreign Economic Policy of the Joint Economic Committee of the Congress of the United States, 91st Congress, 2nd Session, 19 March (Washington, DC: US Government Printing Office).

Johnson, D.G. (1991), *World Agriculture in Disarray*, 2nd edn (New York: St. Martin's Press).

Josling, T.E. (1977a), 'Government price policies and the structure of international agricultural trade', *Journal of Agricultural Economics*, vol. 28, no. 3, pp. 261–77.

Josling, T.E. (1977b), *Agriculture in the Tokyo Round Negotiations* (London: Trade Policy Research Centre).

Josling, T.E. (1993), 'Agriculture in a world of trading blocs', *Australian Journal of Agricultural Economics*, vol. 37, no. 3, pp. 155–79.

Josling, T.E. et al. (1989), *Tariffication and Rebalancing*, Commissioned Paper 4 (St. Paul, MN: International Agricultural Trade Research Consortium, University of Minnesota).

Josling, T.E. et al. (1990), *The Comprehensive Proposals for Negotiations in Agriculture*, Commissioned Paper 7 (St Paul, MN: International Agricultural Trade Research Consortium, University of Minnesota).

Josling, T.E. et al. (1994), *The Uruguay Round Agreement on Agriculture: An Evaluation*, Commissioned Paper 9 (St Paul, MN: International Agricultural Trade Research Consortium, University of Minnesota).

Josling, T.E. and S. Tangermann (1995), 'Tariffication in the Uruguay Round Agreement on Agriculture: Its Significance for Europe', in: R. Gray, T. Becker and A. Schmitz (eds), *World Agriculture in a Post-GATT Environment* (Saskatoon: University Extension Press).

Kelly, W.B. (ed.) (1963), *Studies in United States Commercial Policy* (Raleigh: University of North Carolina Press).

Keohane, R.O. and J.S. Nye (1973), 'World politics and the international economic system', in: C.F. Bergsten (ed.), *The Future of the International Economic Order: An Agenda for Research* (Lexington: Lexington Books).

Kock, K. (1969), *International Trade Policy and the GATT 1947–1967* (Stockholm: Almqvist and Wiksell).

Koenig, E. (1975), 'Agriculture and the MTN', *Foreign Agriculture* (US Department of Agriculture), vol. 13, no. 50, pp. 9–10.

Koester, U. et al. (1988), *Disharmonies in EC and US Agricultural Policy*

Measures, Study prepared for the Commission of the European Communities (Luxembourg: EC Commission).

Kramer, C.S. (1988), *Harmonizing Health and Sanitary Standards in the GATT: Proposals and Issues*, Discussion Paper FAP 88–02 (Washington, DC: National Center for Food and Agricultural Policy).

Leddy, J.M. (1963), 'The United States commercial policy and the domestic Farm Program', in: W.B. Kelly (ed.), *Studies in United States Commercial Policy* (Raleigh: University of North Carolina Press).

Lewis, J.N. (1962), 'The French Plan: Blueprint for world trade without tears', *Review of Marketing and Agricultural Economics*, vol. 30, no. 3, pp. 143–54.

Magiera, S. *et al.* (1990), *Reinstrumentation of Agricultural Policies*, Commissioned Paper 6 (St Paul, MN: International Agricultural Trade Research Consortium, University of Minnesota).

Malmgren, H.B. (1974), 'Techniques and modalities of agricultural negotiations', in: D.G. Johnson and J.A. Schnittker (eds), *US Agriculture in a World Context* (New York: Praeger).

Malve, P. (1972), *For the Development of Dynamic Agricultural Cooperation between the United States and Europe through Negotiation of a New Type of International Commodity Agreement*, Address to the US National Association of Wheat Growers, mimeo, Denver, Colorado, 2 January.

May, B. (1994), *Die Uruguay-Runde: Verhandlungsmarathon verhindert trilateralen Handelskrieg*, Arbeitspapiere zur Internationalen Politik 86 (Bonn: Europa Union).

McFadzean, F. et al. (1972), *Towards an Open World Economy* (London: Macmillan for Trade Policy Research Centre).

McRae, D.M. and J.C. Thomas (1983), The GATT and multilateral treaty making: The Tokyo Round', *American Journal of International Law*, vol. 77, no. 1, pp. 51–83.

Meinheit, E. (1995), *Handelspolitik als Umweltpolitik im Agrarbereich? Eine Ökonomische und GATT-Rechtliche Analyse* (Kiel: Wissenschaftsverlag Vauk).

Merciai, P. (1981), 'Safeguard measures in GATT', *Journal of World Trade Law*, vol. 15, no. 1, pp. 41–65.

Moyer, H.W. and T.E. Josling (1990), *Agricultural Policy Reform: Politics and Process in the EC and the USA* (Ames: Iowa State University Press; New York: Harvester Wheatsheaf).

National Center for Food and Agricultural Policy (1988), *Mutual Disarmament in World Agriculture: A Declaration on Agricultural Trade* (Washington, DC: Resources for the Future).

National Planning Association (1971), *US Foreign Economic Policy in the 1970s* (Washington, DC: National Planning Association).

OECD (1961), *Trends in Agricultural Policies since 1955* (Paris: OECD).

OECD (1967), *Agricultural Policies in 1966* (Paris: OECD).

OECD (1970), *Essays in Honour of Thorkil Kristensen* (Paris: OECD).

OECD (1972), *Policy Perspectives for International Trade and Economic Relations* (Paris: OECD).

OECD (1982a), *Problems of Agricultural Trade* (Paris: OECD).

OECD (1982b), *Minutes of the Council at Ministerial Level as adopted on 10th–11th May 1982*, C(82)58 (Final).

OECD (1987), *National Policies and Agricultural Trade* (Paris: OECD).

OECD (1988), *Agricultural Policies, Markets and Trade* (Paris: OECD).

OECD (1994a), *Review of Agricultural Policies: Hungary* (Paris: OECD).

OECD (1994b), *The New World Trading System: Readings* (Paris: OECD).

OECD (1995), *Review of Agricultural Policies: Poland* (Paris: OECD).

OEEC (1956–7), *Agricultural Policies in Europe and North America: Three reports of the Ministerial Committee for Agriculture and Food* (Paris: OEEC).

OEEC (1960), *Problems in Dairy Policy* (Paris: OEEC).

Office of Economics (1994), *Effects of the Uruguay Round Agreement on US Agricultural Commodities* (Washington, DC: Economic Research Office, US Department of Agriculture).

Ostry, S. (1990), *Governments and Corporations in a Shrinking World* (New York: Council on Foreign Relations Press).

Oxley, A. (1990), *The Challenge of Free Trade*, Report to the Eminent Persons Group on World Trade (Hemel Hempstead: Harvester Wheatsheaf).

Patterson, G. (1966), *Discrimination in International Trade: The Policy Issues 1945–1965* (Princeton: Princeton University Press).

Patterson, G. (1971), 'Current GATT work on trade barriers', in: Williams Commission, *United States International Economic Policy in an Interdependent World*, Report to the President submitted by the Presidential Commission on International Trade and Investment Policy (Washington, DC: US Government Printing Office).

Patterson, G. (1983), 'The European Community as a threat to the system', in: W.R. Cline (ed.), *Trade Policy in the 1980s* (Washington, DC: Institute for International Economics).

Penrose, E.F. (1953), *Economic Planning for Peace* (Princeton: Princeton University Press).

Pescatore, P., W.J. Davey and A.F. Lowenfeld (1991), *Handbook of GATT Dispute Settlement* (Ardsley-on-Hudson: Transnational Juris Publications; Deventer: Kluwer Law and Taxation).

Petersen, P.G. (1971), *A Foreign Economic Perspective*, Council on International Economic Policy (Washington, DC: Executive Office of the President).

Petrey, L.A. and R.W.M. Johnson (1993), 'Agriculture in the Uruguay Round: Sanitary and phytosanitary measures', *Review of Marketing and Agricultural Economics*, vol. 61, pp. 433–42.

Preeg, E.H. (1970), *Traders and Diplomats: An Analysis of the Kennedy Round of Trade Negotiations under the General Agreement on Tariffs and Trade* (Washington, DC: Brookings Institution).

Roningen, V.O. and P.M. Dixit (1990), 'Assessing the implications of freer agricultural trade', *Food Policy*, vol. 15, no. 1, pp. 67–75.

Rosen, J.F. (1989), 'The US–Israel Free Trade Area Agreement: How well is it working and what have we learned?, in: J.J. Schott (ed.), *Free Trade Areas and US Trade Policy* (Washington, DC: Institute for International Economics).

Roth, W.M. (1969), *Future United States Foreign Trade Policy*, Report to the President submitted by the Special Representative for Trade Negotiations, William M. Roth, 14 January (Washington, DC: US Government Printing Office).

Schnittker, J.A. (1970a), in: *A Foreign Economic Policy for the 1970s*, Hearings before the Subcommittee on Foreign Economic Policy of the Joint Economic Committee of the Congress of the United States, 91st Congress, 2nd Session, 19 March (Washington, DC: US Government Printing Office).

Schnittker, J.A. (1970b), 'Reflections on trade in agriculture', in: OECD, *Essays in Honour of Thorkil Kristensen* (Paris: OECD).

Schott, J. (1994), *The Uruguay Round: An Assessment* (Washington, DC: Institute for International Economics).

Scott, B. (1984), *Has the Cavalry Arrived? A Report on Trade Liberalisation and Economic Recovery* (London: Trade Policy Research Centre).

Senti, R. (1994), *GATT–WTO: Die neue Welthandelsordnung nach der Uruguay-Runde* (Zürich: Institut für Wirtschaftsforschung der ETH Zürich).

Tangermann, S. (1994), *An Assessment of the Uruguay Round Agreement on Agriculture*, Paper prepared for the Directorate for Food, Agriculture and Fisheries and the Trade Directorate of OECD, Stanford, June.

Tangermann, S. (1995), *Implementation of the Uruguay Round Agreement on Agriculture by Major Developed Countries*, Report prepared for the United Nations Conference on Trade and Development, UNCTAD Document ITD/16, 3 October.

Tangermann, S., and T.E. Josling (1994), 'The GATT and Community preference for cereals', *Agra Europe*, no. 1602, 15 July, pp. E/8–10.

Tangermann, S., T.E. Josling and S. Pearson (1987), 'Multilateral negotiations on farm-support levels', *The World Economy*, vol. 10, no. 3, pp. 265–81.

't Hooft-Welwars, M.J. (1965), 'The organization of international markets for primary commodities: The French proposals regarding market organization', *Trade and Development, Commodity Trade* series (Geneva: UN), E/CONF. 46/141, vol. 3, pp. 459–64.

Trebilcock, M.J., R. Howse and M.A. Chandler (1990), *Trade and Transitions: A Comparative Analysis of Adjustment Policies* (London and New York: Routledge).

Tumlir, J. (1985), *Protectionism: Trade Policy in Democratic Societies* (Washington, DC: American Enterprise Institute for Public Policy Research).

Tyers, R. and K. Anderson (1988), 'Liberalizing OECD agricultural policies in the Uruguay Round: Effects on trade and welfare', *Journal of Agricultural Economics*, vol. 39, no. 2, pp. 197–216.

UN (1946a), *ECOSOC Res. 13*, Document E/22.

UN (1946b), *Report of the First Session of the Preparatory Committee of the United Nations Conference on Trade and Employment*, Document EPCT/33 (Reprinted as *Preliminary Draft Charter for the International Trade Organization*, US Department of State, Pub. no. 2728, Commercial Policy Series 98, 1947).

UN (1947a), *Report of the Drafting Committee of the Preparatory Committee of the United Nations Conference on Trade and Employment*, Document EPCT/34.

UN (1947b), *Report of the Second Session of the Preparatory Committee of the United Nations Conference on Trade and Employment*, Document EPCT/186 (Reprinted as *Draft Charter for the International Trade Organization of the United Nations*, US Department of State, Pub. no. 2927, Commercial Policy Series 106, 1947).

UN (1947c), *ECOSOC Res. 30*, Document E/437, 28 March.

UN (1948), *Final Act and Related Documents of the United Nations Conference on Trade and Employment*, Document E/Conf.2/78 (Reprinted as *Havana Charter for an International Trade Organization*, US Department of State, Pub. no. 3117, Commercial Policy Series 113, 1948).

UN (1975), *Report of the World Food Conference*, Document E/CONF, 65/20.

UN (1992), 'Rio Declaration', in: *Report of the United Nations Conference on Environment and Development*, Document A/CONF. 151/26 (vol. 1) Annex I, 12 August, pp. 8–13.

US Congress (1970), *A Foreign Economic Policy for the 1970s*, Hearings before the Subcommittee on Foreign Economic Policy of the Joint Economic Committee of the Congress of the United States, 91st Congress, 2nd Session (Washington, DC: US Government Printing Office).

US Department of Agriculture (1967), *Report on the Agricultural Trade Negotiations in the Kennedy Round*, FAS-M-193 (Washington, DC: Foreign Agricultural Service).

US Department of Agriculture (1981), *Report on Agricultural Concessions in the Multilateral Trade Negotiations*, FAS-M-301 (Washington DC: Foreign Agricultural Service).

US Department of Agriculture (1994), *PS&D View*, Database, electronic version (Washington, DC: US Department of Agriculture, Economic Research Service).

US Department of Agriculture (1995), *Foreign Agricultural Trade of the United States*, Fiscal Year 1994 Supplement (Washington, DC: US Department of Agriculture).

US Department of State (1945), Pub. no. 2411, Commercial Policy Series 79, December.

US Department of State (1946), Pub. no. 2598, Commercial Policy Series 93, September.

US Department of State (1964), *Bulletin*, 11 May.

Warley, T.K. (1967a), 'Problems of world trade in agricultural products', in: T.K. Warley (ed.), *Agricultural Producers and Their Markets* (Oxford: Blackwell).

Warley, T.K. (ed.) (1967b), *Agricultural Producers and Their Markets* (Oxford: Blackwell).

Warley, T.K. (1976), 'Western trade in agricultural products', in: A. Shonfield (ed.), *International Economic Relations of the Western World 1959–1971*, vol. I, (London: Oxford University Press for Royal Institute for International Affairs).

Warley, T.K. (1977), *Agriculture in an Interdependent World: US and Canadian Perspectives* (Washington, DC: National Planning Association).

Warley, T.K. (1978), 'What chance has agriculture in the Tokyo Round?' *World Economy*, vol. 1, no. 2, pp. 177–94.

White, E.W. (1972), *Negotiating on Agriculture*, Draft of a paper sent to Hugh Corbet on 8 June.

Wilcox, C. (1949), *A Charter for World Trade* (New York: Macmillan).

Wilgress, D.L. (1948), *Report to the Secretary of State for External Affairs of the Canadian Delegation to the United Nations Conference on Trade and Employment at Havana*, Ottawa, 13 July, p. 30.

Williams Commission (1971), *United States International Economic Policy in an Interdependent World*, Report to the President submitted by the Presidential Commission on International Trade and Investment Policy (Washington DC: US Government Printing Office).

Winham, G.R. (1986), *International Trade and the Tokyo Round Negotiation* (Princeton: Princeton University Press).

Winterling, H.-J. (1986), 'Selbstbeschränkungsabkommen im Internationalen Agrarhandel: Eine Qualitative sowie Quantitative Analyse ihrer Bedeutung und Wirkungen am Beispiel des Tapiokaabkommens zwischen der Europäischen Gemeinschaft und Thailand', *Agrarwirtschaft*, Sonderheft 111.

Wirth, D.A. (1994), *The Role of Science in the Uruguay Round and NAFTA Trade Disciplines*, Environment and Trade No. 8 (Geneva: United Nations Environment Programme).

Wolff, A.W. (1983), 'Need for new GATT rules to govern safeguard actions', in: W.R. Cline (ed.), *Trade Policy in the 1980s* (Washington, DC: Institute for International Economics).

Wolter, F. (1994), *Statement on the Occasion of the 31st General Conference on the International Federation of Agricultural Producers*, mimeo, Istanbul, 3 May.

Wonnacott, P. (1987), *The United States and Canada: The Quest for Free Trade* (Washington, DC: Institute for International Economics).

Worthington, H. (1969), 'Recent efforts of the United States to facilitate the expansion of agricultural trade', in: Agricultural Policy Institute, *Agriculture and International Trade* (Raleigh: Agricultural Policy Institute).

Worthington, H. (1971), 'Special trade negotiating problems for agriculture', in: Williams Commission, *United States International Economic Policy in an Interdependent World*, Report to the President submitted by the Presidential Commission on International Trade and Investment Policy (Washington, DC: US Government Printing Office).

Worthington, H. (1972), 'The environment of the world's trade and monetary system', *Foreign Agriculture* (US Department of Agriculture) vol. 10, no. 11, 13 March, pp. 7–9, 16.

WTO (1995a), *The Results of the Uruguay Round of Multilateral Trade Negotiations: The Legal Texts* (Geneva: WTO) (first published in June 1994 by the GATT Secretariat).

WTO (1995b), *Regionalism and the World Trading System* (Geneva: WTO).

Zaglits, O. (1967), 'Agricultural trade and trade policy', *Foreign Trade and Trade Policy*, vol. 6 (Washington, DC: National Advisory Commission on Food and Fiber).

Zeitz, J. and A. Valdés (1988), *Agriculture in the GATT: An Analysis of Alternative Approaches to Reform*, Research Report 70 (Washington, DC: International Food Policy Research Institute).

Index

286